Biological Techniques Series
Alexander Hollaender, *Editor*

Autoradiographic Techniques: Localization of Radioisotopes in Biological Material
WILLIAM D. GUDE

Introduction to Research in Ultraviolet Photobiology
JOHN JAGGER

The Laboratory Mouse: Selection and Management
M. L. SIMMONS AND J. O. BRICK

THE LABORATORY MOUSE

Selection and Management

M. L. Simmons

*Smith Kline & French Laboratories
Philadelphia, Pennsylvania*

J. O. Brick

*Biology Division, Oak Ridge National Laboratory
Oak Ridge, Tennessee*

Prentice-Hall, Inc., Englewood Cliffs, N. J.

PRENTICE-HALL INTERNATIONAL, INC., *London*
PRENTICE-HALL OF AUSTRALIA, PTY. LTD., *Sydney*
PRENTICE-HALL OF CANADA, LTD., *Toronto*
PRENTICE-HALL OF INDIA PRIVATE LTD., *New Delhi*
PRENTICE-HALL OF JAPAN, INC., *Tokyo*

© 1970 by Prentice-Hall, Inc., Englewood Cliffs, New Jersey

All rights reserved. No part of this book may be reproduced in any form or by any means without permission in writing from the publisher.

Current printing (last digit):
10 9 8 7 6 5 4 3 2 1

13-519876-3

Library of Congress Catalog Number 75-93526

Printed in the United States of America

To: Charles C Congdon, M.D., Head, Mammalian Recovery Group, Biology Division, Oak Ridge National Laboratory, *for his vision into and unfailing support of the modern concepts of laboratory animal care and research.*

Foreword

In the list of modern laboratory techniques the mouse has become an excellent scientific "tool." Formerly, in working with the mouse one had to be satisfied with variable results since little was known about its background, and it was difficult to maintain mice under uniform conditions. This has been changed by new developments.

The mouse has a combination of advantages over other mammals for certain areas of research:

(1) Its genetics is fairly well understood.

(2) Inbred and hybrid strains are available, often with specific and unique characteristics.

(3) The maintenance of colonies of large numbers of highly uniform, healthy mice has been made possible by advances such as filter top techniques and germfree animals, which are described in this book.

(4) The relatively low cost of handling large numbers of individual mice has made the mouse economically feasible as a laboratory tool.

This book describes the most up-to-date procedures as well as the things that one should guard against, such as diseases, bacterial infection, viruses and other parasites. In addition, it explains how to keep the mouse clean and healthy.

A good part of the book is based on experiences in the Biology Division of Oak Ridge National Laboratory where highly successful animal facilities have been developed. Excellent control of the genetic makeup of the mouse and a minimum of disease problems are present in these facilities.

I believe that this book will be useful for scientists who are interested in using the mouse as a scientific tool.

ALEXANDER HOLLAENDER

Preface

One of the most important factors in biological research involving mammalian systems is the experimental animal used. The mouse has been used extensively for many years, partly because it survives under unfavorable environmental conditions, and partly because it reproduces and matures rapidly enough to ensure that successive generations are easily and quickly available either for genetic evaluations or simply in adequate numbers for specific experimental designs. Of major importance, of course, are the economic advantages ensuing from its small size.

This book introduces the student and researcher with little or no previous experience in handling mice to considerations that are necessary for efficient use of this animal in the laboratory. The species of mouse that we refer to most of the time is *Mus musculus*, primarily because it is more often used in research than any other. A table (Table 3.1) has been prepared which includes the location of the mouse within the animal kingdom and a compilation of data prepared from various sources. All of the information in the table is directly applicable to the laboratory mouse and much of what is written applies to *any* mouse, indeed to other mammals (*e.g.*, the rat) as well.

Going beyond a mere treatment of problems involved in the day-to-day care of laboratory mice, the text describes current methods, equipment, and technology, and suggests reference sources for specific information on related subjects. We have traced the historical development of the mouse as a research animal, recorded general operating procedures for use of the mouse in the laboratory, and discussed important related modern subjects (*e.g.*, the problem of serious allergic reactions induced in humans by close contact with mice).

Although this brief book cannot be comprehensive on so broad a subject, it should nonetheless help to fill many needs. In short, it should be useful to the student, to the scientist who needs a ready reference source, and to the researcher who can utilize the information available to properly conduct experiments involving mice.

We acknowledge and thank Dr. Alexander Hollaender, Senior Research Advisor, Biology Division, Oak Ridge National Laboratory, Dr. L. H. Smith, Dr. C. B. Richter, Dr. R. W. Tennant, and Mr. John Lenz, for their encouragement, suggestions, and contributions, and Mr. Craig Whitmire, Jr., and Mrs. Anne Skeel of the Biology Division Editorial Department, for their help in preparation of the manuscript. We also wish to thank the many contributors of photographs and techniques that form a valuable part of this book.

M. L. SIMMONS AND J. O. BRICK
Oak Ridge, Tennessee

Contents

	Foreword	vii
	Preface	ix
ONE	**Ecological Classification of the Mouse**	
	Introduction	1
	Specific Ecological Classification	2
	Germfree Mice	2
	Defined-Flora Mice	6
	Specific-Pathogen-Free Mice	7
	Conventional Mice	11
TWO	**Materials and Methods for Breeding and Housing Laboratory Mice**	
	Introduction	16
	Animal Caging Systems	17
	Bedding	24
	Bedding Disposal	25
	Breeding	26
	Systems	27

Housing	28
Conventional	28
Germfree, Defined - Flora, and Specific-Pathogen-Free	29

THREE   Animal Health

Introduction	49
Testing and Monitoring Programs	50
Disease Prevention	52
Nutritional Requirements	53
Disease Control	54
Disease in Laboratory Mice	57
Noninfectious Disease	59
Infectious Disease	62
Parasites Infecting Mice	95
Experimental Disease	101
Genetic Disease	102
Neoplastic Disease	103
Nutritional Deficiencies	106
Zoonoses and Allergy	110
Zoonoses	110
Allergy to Laboratory Mice	113

FOUR   Techniques

Introduction	125
Surgical Procedures on the Mouse	125
Preparation	126
Surgery	126
Postsurgical Care	126
Fluid Administration	127
Intraperitoneal	127
Intramuscular	127
Subcutaneous	128
Intrathoracic	128
Intravenous	128
Intra-arterial	129
Oral Administration	129

Methods of Blood Withdrawal	129
Orbital Sinus	129
Decapitation	129
Cardiac Puncture	130
Lymph Collection	130
Ascites Fluid Collection	130
Urine Collection	131
Special Techniques	131
Drug Administration and Dosage	136
Anesthetics	137
Injectable Anesthetics	137
Inhalation Anesthetics	140
Other Types	143
Euthanasia	143
Roentgenographic Technique	144
Radioisotopes	145

FIVE   Technological Advances in Laboratory Animal Management

Introduction	149
Hypoxia Chamber	150
Filter-Cage Systems	153
Computerized Data Storage and Retrieval	156
Automatic Watering Systems	159

Appendix

A. Manufacturers	165
B. Sources of Laboratory Mice	168
C. Suggested Library	169
D. Organizations as Sources of Information	172

Index   173

THE LABORATORY MOUSE

Selection and Management

Ecological Classification of the Mouse

ONE

INTRODUCTION

Recognizing that the more uniform the experimental animal the fewer the number necessary to attain a given standard of accuracy or repeatability, research workers in the biological sciences have developed the mouse into one of the most refined components of the experimentalist's basic equipment. Through years of research, numerous useful strains and types have become available. To reduce variables caused by *genetic* differences, scientists have developed inbred and hybrid colonies; more recently, to counteract errors induced by *environmental* differences, they have introduced *ecological classifications*.

Ecological classification can be defined as the relationship of the mouse to its particular and specific environment. This environment includes the physical surroundings, the organisms associated with the mouse, and the organisms present within the limits of the physical surroundings. In Chapter 1, we define and discuss four such classifications—germfree (GF), defined-flora (DF), specific-pathogen-free (SPF), and conventional. Related terminology, derivation and technological techniques, and applications of each classification are also discussed.

SPECIFIC ECOLOGICAL CLASSIFICATION

Germfree Mice

Definition: Germfree mice are free from all *detectable* internal and external parasites, bacteria, yeast, fungi, protozoa, algae, rickettsia, and viruses. A similar term often used in this context is *axenic*, meaning "free from strangers."

Pasteur's theory that microorganisms are essential for higher forms of life was refuted by COHENDY (1912) when he was successful in raising germfree chickens for periods of up to 40 days. It was not until 1948, however, that REYNIERS *et al.*, using bantam chickens, were successful in demonstrating that reproduction was possible in a germfree system.

In the interim, the technology of germfree research has advanced tremendously. Plastic materials, replacing the much more cumbersome and expensive steel isolators, have been a significant and dramatic improvement. Examples of improved isolators and housing used are described in detail in Chapter 2. Many different types of germfree animals have been successfully raised including rats, mice, guinea pigs, rabbits, chickens, dogs, pigs, and lambs. Although some of these have not reproduced in a germfree environment, rats and mice do so quite well.

In the pregnant mouse, the developing embryos are protected from contamination by the placental barrier, a semipermeable membrane made up of placental tissues and limiting the character and amount of material exchanged between mother and fetus in the uterus. Within this protected environment the young are essentially noncontaminated, but following birth, they are usually exposed randomly to various organisms. To derive germfree mice, the problem is to preserve the virtually noncontaminated state *after* the mice lose the protection of the placental barrier.

Recent technological advances have provided a workable method—the sterile hysterectomy of the full-term gravid uterus and its subsequent introduction into a sterile isolation facility. For the offspring to survive, the uterus should be removed from the pregnant mouse within the 24-hr period prior to normal birth. The gestation period varies with each strain, but in general 19 to 21 days is the usual span for the mouse. After the mother is killed by cervical fracture, the ventral abdomen is prepared for surgery. First the abdomen is shaved; then the female's body is immersed in a suitable sterilizing solution at body temperature and pinned on a sterile cloth, the abdomen covered with an adhesive sterile plastic drape.

A midline incision is made through the plastic drape and skin. The intact gravid uterus is aseptically removed after the ovarian ligaments are cut free and the uterus is incised just posterior to the cervix. Should contam-

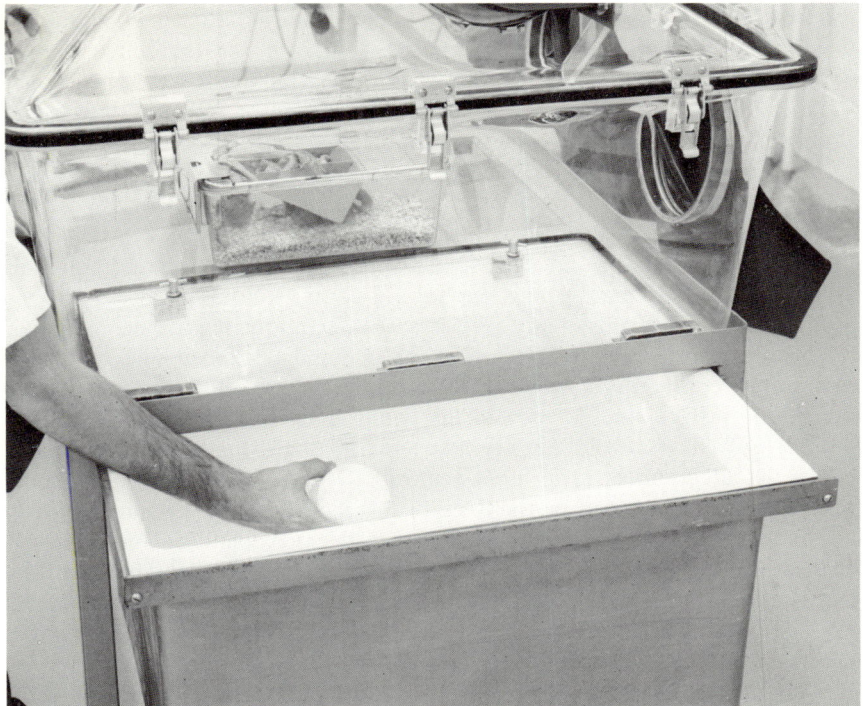

Fig. 1-1. Aseptic transfer of gravid uterus into sterile isolator.

ination be suspected at any time during any part of the process, the entire procedure must be started over with a new female. Additional sterile equipment must be available at all times in the event that a second start is necessary.

The gravid uterus must be protected against temperature shock by placing it inside a tightly sealed container in a warm disinfectant solution. This entire unit is immediately passed into a sterile isolator (Fig. 1.1).

Once inside the isolator, a second person receives the sealed container, removes and opens the gravid uterus, and then carefully removes each pup. Throughout this transfer process, as in the operation, aseptic techniques are necessary. Again, it is very important to keep the fetuses warm, dry, and attached to the placenta for a short period of time after removal. At this point, with the newborn pups aseptically delivered into a sterile isolator, the investigator has surgically derived germfree mice and has the option of using either the "foster-nurse" or "hand-rearing" system to keep them germfree.

In the foster-nurse system, the young can be nursed by an already germfree lactating female inside the isolator. Closely matched pregnancies inside and outside the isolator are required to assure that the delivery date of the foster mother is slightly in advance of that of the female outside so that the nurse mouse is in full lactation at the time of transfer. It is very important to make sure that the foster mother accepts the offspring and is not disturbed for several days while adjusting to them (BLEBY, 1967). Ether or some other chemical is often used to deaden the foster mother's sense of smell so that she will accept the new mice as her own.

If no germfree nurse mice are available, it is necessary to hand rear the surgically acquired young. (This method, of course, must be used to obtain the first germfree animals of any species.) This technique is extremely tedious and not always successful. The problems of hand rearing young mice are frustrating: contamination is more difficult to prevent, urination and defecation must be stimulated, and the young must be fed every two hours. The work of PLEASANTS (1959) illustrates the hand-rearing method in detail.

In actual practice, little hand rearing is done. Most new colonies of germfree mice are derived by the foster-nurse method. Germfree nurse mice can be purchased through commercial breeders (Appendix) and used in the derivations of germfree and other ecological classifications. The germfree state is a first step in attaining defined-flora and specific-pathogen-free mice.

The highly sophisticated monitoring system necessary to determine the actual continuing ecological status of mouse colonies is described in Chapter 3.

The proper diet and aseptic introduction of feed into the isolator is of paramount importance in maintaining a germfree system. Many failures occur in germfree operations as a result of unsterile feed being introduced into the isolator. Oversterilization of feed can result in additional problems, mainly those associated with nutritional deficiencies. Symptoms of specific nutritional deficiencies are discussed in more detail in Chapter 3.

In defining germfree mice at the start of this section, we emphasized the word *detectable*. The point to be made is that *any* ecological classification is only as good as the diagnostic evidence upon which the classification is based. The fact that some bacteria and many viruses are difficult to culture or, in fact, to demonstrate at all, limits the degree to which an animal can be called "germfree." Also, the operative procedure for deriving germfree mice is not foolproof. Unfortunately, the placental barrier does not block all contamination; some viruses, for example, are known to be transmitted through the placental membranes. An example of this sort of intrauterine or vertical transmission is the lymphocytic choriomeningitis virus. According to TRAUB (1936, 1939), vertical transmission may become the primary mode of infection of this virus in a stabilized colony.

More recently, POLLARD (1966), speaking at the Symposium on Viruses of Laboratory Rodents, indicated that leukemogenic agents had entered a germfree barrier by either one of two methods—passage with germinal plasm or through the placenta—but he was uncertain which route was involved.

If germfree mice can be derived from existing colonies that have been tested and are known to be free of agents that can be transmitted vertically, the chances would appear to be improved for reaching a true germfree status. At any rate, to classify mice as "germfree," we must test for a wide variety of organisms in order to know that the animals are actually "germfree."

Germfree animals are *not* antigen-free animals. As long as killed-but-whole organisms are in feed, bedding, and water that are directly associated with the animals, the mice will be subjected to numerous antigenic stimulations. The complex subject of an antigen-free environment is beyond the intent of this book.

Despite these limitations, however, enough information has accumulated to make germfree animals an additional research tool in the study of cancer, radiation immunology, and enteric, dental, and nutritional diseases. A discussion of potential applications of germfree animals is given by MICKELSEN (1962). For many examples of germfree application in research ranging from silkworm production to lymphatic tissue responses in mice and rats, we suggest reviewing the Proceedings of the International Symposium on Germfree Life Research (held in Nagoya, Japan in 1967).

Another commonly used word in the literature is "gnotobiote" (known life). Such an animal, as the term is generally used, could be either germfree or contaminated with one or more demonstrable organisms; the criterion is that the animal's ecological classification be *known*. In this sense, a "germfree" animal becomes a "gnotobiote," whereas a "gnotobiote" may or may not be germfree. Similarly, mice described under the ecological title of "defined-flora" are also properly referred to as gnotobiotes because each associated organism is known, at least by the latest methods available to test for it.

Different levels of gnotobiotes are recognized and are often described as monocontaminant (one organism) and polycontaminant (more than one organism). To avoid conflicts in terminology, we will restrict our discussions to four ecological classes of mice (i.e., germfree, defined-flora, specific-pathogen-free, and conventional).

Recommended Reading on Germfree Mice

Cumming, C. N. Wentworth. "Large-Scale Production of Mice and Rats in a Controlled Environment," in *Husbandry of Laboratory Animals*, 3rd Symposium of the International Committee on Laboratory Animals, ed. by M. L. Conalty (New York: Academic Press, 1967), pp. 51–60.

Dinsley, Marjorie. "Germfree Animals," in *The UFAW Handbook on the Care and Management of Laboratory Animals*, 3rd ed., ed. by W. Lane-Petter *et al.* (Baltimore: The Williams & Wilkins Co., 1967), pp. 216–236.

Lev, M. "Germ-Free Animals," in *Animals for Research: Principles of Breeding and Management*, ed. by W. Lane-Petter (New York: Academic Press, 1963), pp. 139–175.

Luckey, Thomas D. *Germfree Life and Gnotobiology* (New York: Academic Press, 1963), 512 pp.

Miyakawa, M., and T. D. Luckey, eds. *Advances in Germfree Research and Gnotobiology* (Cleveland: The Chemical Rubber Co. Press, 1968), 439 pp.

Peters, Arthur C., and John H. Litchfield. "Germfree Animals: Their Application in Biological Research." *Battelle Tech. Rev.* **12**: 9–14, 1963.

"Proceedings of the Third Symposium on Gnotobiotic Technology." *Lab. Animal Care* **13**: 569–679, 1963.

Defined-Flora Mice

Definition: Defined-flora mice are germfree animals that have been *intentionally* contaminated with specific microorganisms or parasites, and then continuously monitored to assure the continued presence of these selected organisms and the absence of all others.

Clarence Cook Little started the first inbred strain of mice in 1909. Since that time, the pursuit of inbred breeding systems has led to almost complete genetic control of the mouse. The contemporary development of germfree techniques has provided another step toward control of the laboratory mouse. By controlling the microorganisms present in the mouse, long-term experiments can be conducted with research mice that are microbially as well as genetically alike throughout the experiment. With a system that provides a constant supply of microorganisms from a central source, such as the one proposed by BAKER (1966), many experiments could be repeated or expanded by other investigators using genetically and microbially stable mice.

The first step in deriving defined-flora mice is to obtain germfree animals (either through commercial suppliers or via the procedures described earlier). Once germfree status is achieved and diagnostically confirmed, any number of preselected microorganisms can be administered in several different ways. One such method requires the transfer of germfree mice from an isolator into a restricted barrier facility. Once inside, the germfree mice may be given food pellets saturated with inoculums of specific bacteria. The mice subsequently born within the barrier facility are self-contaminated with the same bacteria from the feces of the parents.

A second and more reliable procedure is to feed pure cultures of the selected contaminant(s) to the mice before removing them from the holding or transfer isolators. After the selected flora is established, the mice should be tested before being transferred into a barrier facility.

The selection of an ideal flora is always of concern to investigators using defined-flora mice. A considerable amount of research has been devoted to the relationship of the selected enteric flora to the growth rate, susceptibility to experimental infection, and the effect of endotoxins (DUBOS and SCHAEDLER, 1960, 1962; SCHAEDLER and DUBOS, 1962). DUBOS and SCHAEDLER (1964) also pointed out that the indigenous mouse flora consists of two types—(1) organisms that become established, yet do not appear to be necessary for survival, but may become detrimental under certain circumstances; (2) autochthonous flora, that is, those organisms that exist in a type of symbiotic relationship with host or each other. Work of this nature is presently under way by UPTON et al. in this laboratory.

Specific-Pathogen-Free Mice

Definition: Specific-pathogen-free mice are "free of specified microorganisms and parasites, but not necessarily free of the others not specified" (nomenclature recommended by the International Committee on Laboratory Animals, 1964).

A very significant change is presently taking place in the concept of breeding and maintaining research mice. This change represents many years of research directed toward complete control of the flora and fauna of animals and is, in part, a result of efforts to establish germfree techniques for control of infectious diseases. Within the limits of present viral and bacteriological testing, this control has been achieved in many instances for the laboratory mouse through the development of SPF techniques.

Specific-pathogen-free mice are derived and maintained free from specified contaminants (as opposed to defined-flora mice, which are intentionally exposed to specific contaminants). In actual usage, however, the current trend leans more toward the development of a completely "pathogen-free" animal rather than an animal that is free from only a number of selected pathogens.

Epidemic bacterial diseases of laboratory mice have often been partially controlled through improved management and housing standards. In the past, antibiotics and anthelmintics have been used to prevent epidemics, although testing, accompanied by rigid culling of infected animals, is usually more effective. Freedom from a number of pathogenic parasites (rickettsia, viruses, and fungi) has resulted only after extensive use of both therapeutic and culling methods.

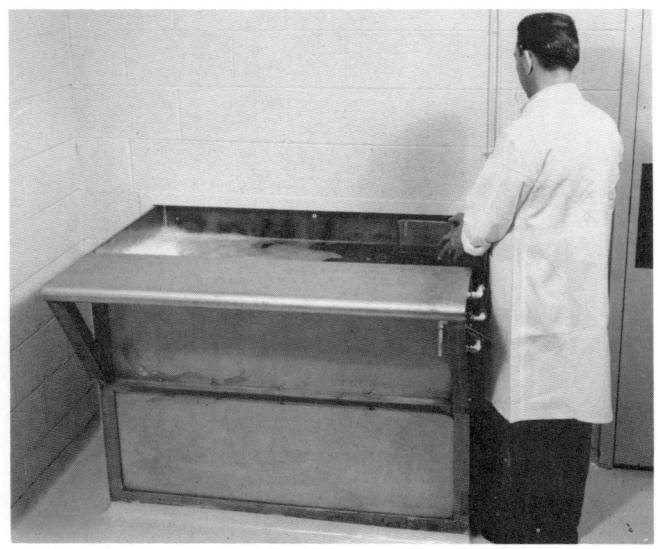
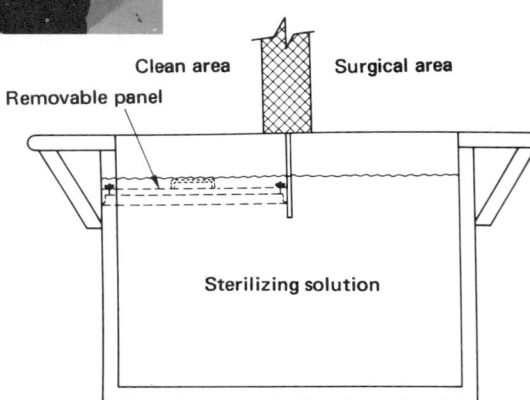

Fig. 1-2. (a) Wall-mounted, pass through dip tank for transfer of mice or supplies into a barrier. (b) Schematic side view of Fig. 1-2a.

To establish a colony of SPF mice, germfree mice are obtained and infected with a known nonpathogenic flora. Then the SPF mouse is housed in a "clean" environment that no longer requires the rigid, sterile conditions of a germfree isolator. Pathogen-free mice and/or material required for maintenance of such mice can be transferred into a barrier through a wall-type dip tank (as shown in Fig. 1.2), such as the one described by SIMMONS et al. (1967).

The germfree mice can be given a known bacterial flora at weaning, or, while in contact with their foster mother, they will eventually establish the foster mother's bacterial flora. Of course, if any pathogens are present, they will also be likely to establish. The colony is continually monitored to assure absence of specified pathogens.

Another method of transferring mice into an SPF barrier facility involves the use of sterile transfer containers. In this manner, previously established

Specific Ecological Classification 9

Fig. 1-3. Transfer of pathogen-free mice in sterile, sealed plastic bags.

Fig. 1-4. Charles River Breeding Laboratory's germfree shipper. (Photo courtesy of Charles River Breeding Laboratory.)

SPF mice can be transferred from one isolation facility to another. BRICK *et al.* (1969) have described a transfer procedure in which filter-top cages inside double sterile plastic bags are used (Fig. 1.3). In addition, commercially used plastic transfer shippers assure complete maintenance of sterility during the transfer procedure (Fig. 1.4).

The pathogen-free mouse has been widely used as a research tool. Reports in the literature indicate primarily the results of comparisons made between this mouse and its conventional counterpart on the basis of viral contamination and pathological lesions. Examination for the presence of polyoma virus, K virus of mice, thymic agent, mouse salivary gland virus, mouse adenovirus, and Reovirus type 3 revealed that only Reovirus type 3 existed in pathogen-free as well as conventional colonies (ROWE et al., 1963). Conventional colonies will usually have at least four latent murine viral contaminants.

In a comparison of nine inbred SPF mouse strains with their conventional counterparts for the presence of Reovirus type 3, Theiler's encephalomyelitis, pneumonia virus of mice, K virus, polyoma, mouse adenovirus, Sendai, and mouse hepatitis, all viruses appeared in conventional mice, with an incidence of between 11 and 37%, whereas only Reovirus type 3 and Theiler's encephalomyelitis virus appeared in SPF mice, with a 3% incidence (VAN HOOSIER et al., 1966). In work reported recently, 967 serum samples from SPF retired breeders were examined for seven murine viruses over a 2.5-year period. Results indicated a 1.2% incidence of Reovirus type 3 titers and a 1.5% incidence of GDVII titers (BRICK et al., 1969). As pointed out by GLEDHILL (1965), LCM virus, Riley agent, and possibly Reovirus type 3 may be able to cross the placental barrier. He goes on to say, however, that using fetuses obtained by hysterectomy and reared in an environment that excludes viruses is an important step toward reduction of carried viruses that do not normally produce disease.

In long-term or aging studies, SPF mice are very useful because of longer-than-normal life-spans (NELSON and COLLINS, 1960). With the concept of raising mice free of certain known murine viruses, experimental work involving specific contamination with a single virus becomes more meaningful from a pathological viewpoint. Specific-pathogen-free mice are more resistant than their conventional counterparts to the effects of certain anesthetics, such as fluothane, chloroform, ether, cyclopropane, and trilene (DAVEY, 1962). Long-term toxicity studies indicate that the results gained from using SPF animals more than justified the establishment of the SPF breeding unit (WALKER and POPPLETON, 1967). As mentioned previously, the validity of any ecological classification is only as good as the diagnostic evidence relating to that freedom. Frequent monitoring of an SPF colony is absolutely necessary to be sure that undesirable contaminants have not become established. How often and how many samples are needed for an awareness of colony status are extremely difficult questions to answer. It is, however, necessary to randomize samples from all types of mice (breeders and experimental) and from all areas within the barrier facility.

The SPF mouse may never completely replace the conventional mouse as a research tool. However, the SPF mouse is being utilized in more and more of the research being conducted today. As the demand for this type of mouse increases, the overall cost of production and holding decreases. Money expended in producing and caring for animals of improved quality becomes a form of insurance for the successful outcome of experiments.

Recommended Reading on SPF Mice

Drepper, K. "Production of Diets for SPF and Germ-Free Animals with Special Regard to Injury of the Protein Value Following Sterilization," in *Husbandry of Laboratory Animals*, 3rd Symposium of the International Committee on Laboratory Animals, ed. by M. L. Conalty (New York: Academic Press, 1967), pp. 207–216.

Foster, H. L. "Establishment and Operation of SPF Colonies," in *The Problems of Laboratory Animal Disease*, ed. by R. J. C. Harris (New York: Academic Press, 1962), pp. 249–259.

Foster, H. L., Sumner J. Foster, and Erving S. Pfau. "The Large Scale Production of Caesarean-Originated, Barrier-Sustained Mice." *Lab. Animal Care* **13**: 711–718, 1963.

Meister, G., H. P. Hobik, and K.-G. Metzger. "Comparative Bacteriological and Pathological Investigations on Commercially Available SPF-Animals," in *Husbandry of Laboratory Animals*, 3rd Symposium of the International Committee on Laboratory Animals, ed. by M. L. Conalty (New York: Academic Press, 1967), pp. 387–391.

Paget, G. E. "The Pathological State of Specific-Pathogen-Free Animals." *Proc. Roy. Soc. Med.* **55**: 262, 1962.

Trentin, John J., G. L. Van Hoosier, Jr., Jacqueline Shields, Kristina Stephens, and Wayne A. Stenback. "Establishment of a Caesarean-Derived, Gnotobiote Foster Nursed Inbred Mouse Colony with Observations on the Control of Pseudomonas." *Lab. Animal Care* **16**: 109–118, 1966.

Trentin, John J., G. L. Van Hoosier, Jr., J. Shields, K. Stephens, W. A. Stenback, and J. C. Parker. "Limiting the Viral Spectrum of the Laboratory Mouse." NCI Monograph 20 (*Viruses of Laboratory Rodents*), 147–157, 1966.

Wheater, D. W. F. "The Bacterial Flora of an SPF Colony of Mice, Rats, and Guinea Pigs," in *Husbandry of Laboratory Animals*, 3rd Symposium of the International Committee on Laboratory Animals, ed. by M. L. Conalty (New York: Academic Press, 1967), pp. 343–360.

Conventional Mice

Definition: All mice that are not germfree, defined-flora, or specific-pathogenfree are arbitrarily called "conventional."

Because of the difficulty in defining "conventional" mice, much misunderstanding has come about in collaborative research among different research centers and also between research facilities and commercial suppliers. Our method of defining this classification is to define all other specific ecological classifications (i.e., germfree, defined-flora, and specific-pathogen-free) and then simply to call the remaining types "conventional." We would prefer to define conventional animals in a less arbitrary way; however, without such an understanding, mice considered conventional in one research facility could be considered by another to be specific-pathogen-free.

Under our definition, mice that are spontaneously infected with *any* of the murine viruses, parasites, bacterial and rickettsial pathogens, and pathogenic protozoa could conceivably be considered conventional, whereas animals infected with all would also be considered conventional. Thus, we have the enigma of some mice being more conventional than others, but still falling within the definition. Nonetheless, keeping such animals in a general classification distinct from germfree, defined-flora, and specific-pathogen-free mice is useful.

As a practical matter, in many instances, choosing the degree of conventionality of the mice depends primarily on two factors: (1) the source(s) of the mice brought into the facility from outside; (2) the conditions under which the mice are to be maintained within the research facility. Obviously, the intended usage of the experimental animals is the key: It does not make much sense to demand infectionfree animals from your suppliers and then bring the animals directly into a highly infected environment. The method of processing incoming animals is also important. Quarantine and testing procedures are necessary when conventional animals are received unless the investigator is willing to accept at random all incoming infections from commercial sources.

Figure 1.5 represents diagrammatically what happens when animals are supplied from both outside and inside sources into a conventional colony. Conventional animals that best suit individual research purposes must be selected (depending on the degree of infection that can be tolerated within the facility), and the testing program and quarantine system must be adjusted accordingly. However, one must test and monitor the inside pool, as well as the incoming animals.

Whereas other ecological classifications require special housing systems, housing for conventional mice varies all the way from woodshed-type facilities to well-lighted and well-ventilated animal space. Of course, this much wider range of environmental conditions can be tolerated with conventional animals because they are already contaminated to some extent.

Most mice used in biological research over the past 20 years have been conventional. They are relatively inexpensive to maintain and are adequate for many experiments. They have been used in practically every type of

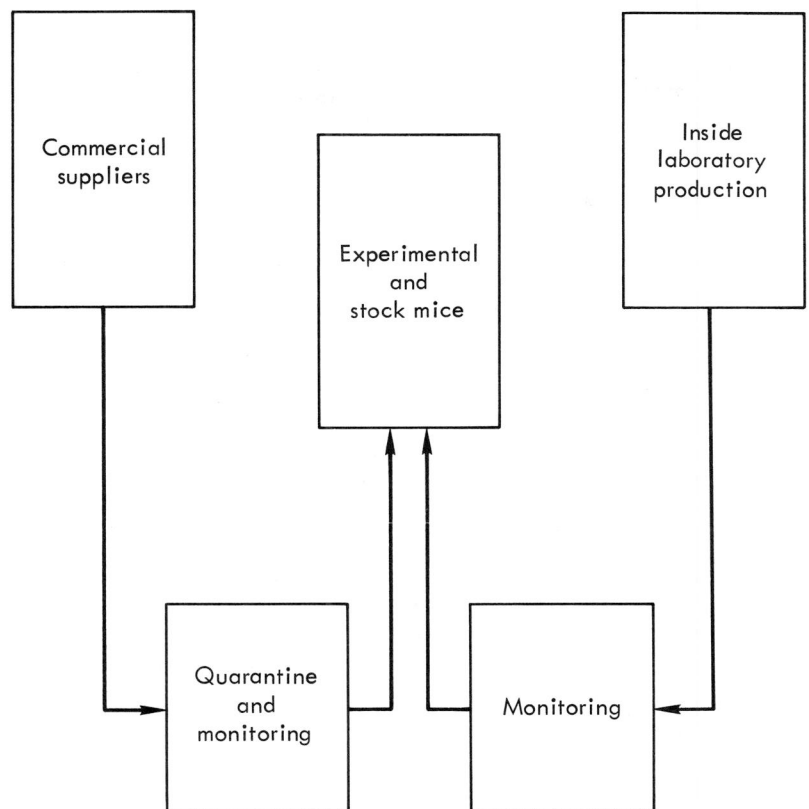

Fig. 1-5. Flow diagram of conventional animal facility sources.

experiment from genetic to surgical. In general, however, when the parameters being studied are fine or the stress factor is great, conventional mice may be inadequate. A classic example would be the early death in lethal and supralethal X-irradiated conventional mice infected with *Pseudomonas aeruginosa* sp. n. (MILLER *et al.*, 1952).

CHAPTER 1 REFERENCES

Baker, Dennis E. J. "The Commercial Production of Mice with a Specified Flora." NCI Monograph 20 (*Viruses of Laboratory Rodents*), 161–172, 1966.

Bleby, John. "Specific-Pathogen-Free Animals," in *The UFAW Handbook on the Care and Management of Laboratory Animals*, 3rd ed., ed. by W. Lane-Petter *et al.* (Baltimore: The Williams & Wilkins Co., 1967), pp. 201–215.

Brick, J. O., R. F. Newell, and D. G. Doherty. "A Barrier System for a Breeding and Experimental Rodent Colony: Description and Operation." *Lab. Animal Care* **19**: 92–97, 1969.

Chance, M. R. A. "The Contribution of Environment to Uniformity: Variance Control, Refinement in Pharmacology." *Laboratory Animals Bureau, Collected Papers*, **6**: 59–73, 1957 (England).

Cohendy, M. "Expériences sur la vie sans microbes." *Ann. Inst. Pasteur* **26**: 106–137, 1912.

Davey, D. G. "The Use of Pathogen Free Animals." *Proc. Roy. Soc. Med.* **55**: 256, 1962.

Dubos, R. J., and R. W. Schaedler. "The Effect of the Intestinal Flora on the Growth Rate of Mice, and on Their Susceptibility to Experimental Infections." *J. Exp. Med.* **111**: 407, 1960.

———. "The Effect of Diet on the Fecal Bacteria Flora of Mice and on Their Resistance to Infection." *J. Exp. Med.* **115**: 1161, 1962.

———. "The Digestive Tract as an Ecosystem." *Am. J. Med. Sci.* **248**: 267–271, 1964.

Gledhill, A. W. "Significance of Pathogen Burden." *Food Cosmet. Toxicol.* **3**: 37, 1965.

"International Committee on Laboratory Animals: Terms and Definitions." *I.C.L.A. Bull. No. 14* (London: I.C.L.A., 1964).

Mickelsen, O. "Nutrition—Germfree Animal Research." *Ann. Rev. Biochem.* **31**: 515–548, 1962.

Miller, C. P., C. W. Hammond, M. Tompkins, and G. Shorter. "The Treatment of Postirradiation Infection with Antibiotics; An Experimental Study on Mice." *J. Lab. Clin. Med.* **39**: 462–479, 1952.

Nelson, J. B., and G. R. Collins. "Establishment and Maintenance of a Specific-Pathogen-Free Colony of Swiss Mice." *Proc. Animal Care Panel* **11**: 65, 1960.

Pleasants, J. R. "Rearing Germfree Caesarean-Born Rats, Mice, and Rabbits Through Weaning." *Ann. N.Y. Acad. Sci.* **78**: 116–126, 1959.

Pollard, Morris. "Viral Status of 'Germfree' Mice." NCI Monograph 20 (*Viruses of Laboratory Rodents*), 167–172, 1966.

Reyniers, J. A., P. C. Trexler, R. F. Ervin, M. Wagner, T. D. Luckey, and H. A. Gordon. "Some Observations on Germfree Bantam Chickens." *Lobund Rep.* **2**: 119–148, 1949.

Rowe, W. P., J. W. Hartley, and R. J. Huebner. "Polyoma and Other Indigenous Mouse Viruses." *Lab. Animal Care* **13**: 166, 1963.

Schaedler, R. W., and R. J. Dubos. "The Fecal Flora of Various Strains of Mice. Its Bearing on Their Susceptibility to Endotoxin." *J. Exp. Med.* **115**: 1149, 1962.

Simmons, M. L., C. B. Richter, R. W. Tennant, and J. A. Franklin. "Production of Specific Pathogen-Free Rats in Plastic Germfree Isolator Rooms," in *Advances in Germfree Research and Gnotobiology*, International Symposium on Germfree Life Research, Nagoya, Japan, April, 1967, ed. by M. Miyakawa and T. D. Luckey (Cleveland: The Chemical Rubber Co. Press, 1968), pp. 38–47.

Simmons, M. L., L. P. Wynns, and E. E. Choat. "A Facility Design for Production of Pathogen-Free, Inbred Mice." *ASHRAE* **9**: 27–31, 1967.

Traub, E. "The Epidemiology of Lymphocytic Choriomeningitis in White Mice." *J. Exp. Med.* **64**: 183–200, 1936.

———. "Epidemiology of Lymphocytic Choriomeningitis in a Mouse Stock Observed for Four Years." *J. Exp. Med.* **69**: 801–817, 1939.

Van Hoosier, G. L., Jr., J. J. Trentin, J. Shields, K. Stephens, W. A. Stenback, and J. C. Parker. "Effect of Caesarean-Derivation, Gnotobiotic Foster Nursing and Barrier Maintenance on an Inbred Mouse Colony on Enzootic Virus Strains." *Lab. Animal Care* **16**: 119, 1966.

Walker, A. I. T., and W. R. A. Poppleton. "The Establishment of a Specific-Pathogen-Free (SPF) Rat and Mouse Breeding Unit." *Lab. Animals* **1**: 1, 1967.

Materials and Methods for Breeding and Housing Laboratory Mice

TWO

INTRODUCTION

Daily care and management of the laboratory mouse and the associated facilities are complicated when we consider different ecological classes of mice. The ecological classification is determined not only by the derivation of the mouse but also by the surrounding environment in which the mouse lives. The degree of contamination to which the mouse is subjected is controlled primarily by the type of facility involved and the procedures (cleaning, showering, sterilizing, etc.) that are carried out within this particular facility.

Chapter 2 is concerned directly with procedures and facility designs used in producing and holding the four types of mice discussed in Chapter 1. Inasmuch as many methods overlap in these four categories, they will not be discussed in detail for each ecological class. However, methods or equipment unique to a specific class are discussed in some detail.

This chapter is not intended to convey all the possible methods of production and maintenance of laboratory mice. It is, however, designed to provide the reader with enough information concerning recent concepts and direct references to evaluate the type of breeding and housing that best suits his particular facility.

ANIMAL CAGING SYSTEMS

Although animal rooms should be designed with the idea of housing only one species at a time, the likelihood of that room continuously housing the same number or same species is remote. The typical animal room will from time to time have cages, cage racks, or animals that cause a variable type disruption of the air flow. In this section, several types of caging systems are discussed. Selection of the best type of cage will depend primarily on the type of work to be done within the facility and on personal preference as related to cost, durability, stackability, and ease of cleaning.

There are many sizes, shapes, and kinds of cages. These range from the older wooden shoebox to glass jars, disposable plastic, single or double nondisposable plastic, and single or double stainless steel boxes. The metal cage may be an open-bottom, suspended type, requiring sliding refuse trays under each row. If the cages are of the solid-bottom type, they may still use suspension-type racks, adjustable solid-shelf racks, ceiling-hung racks, or (least desirable) permanent wall-mounted shelves (Fig. 2.1).

Fig. 2-1. Animal room with permanent wall-mounted shelves.

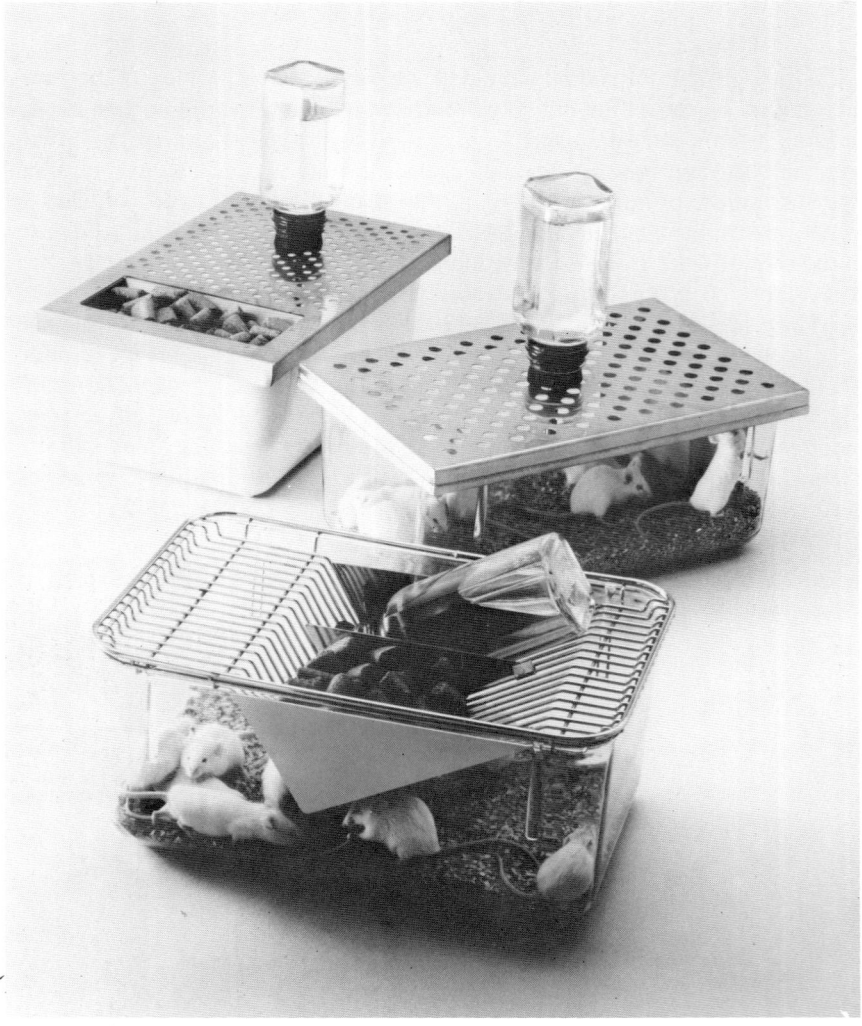

Fig. 2-2. Three types of plastic shoebox mouse cages. (Photo courtesy of Lab-Cages Inc.)

Two of the best mouse cages in use today are the plastic shoebox type (Fig. 2.2) and the stainless steel solid-bottom cage (Fig. 2.3). Sizes vary considerably and several research laboratories have had cages made to their exact specifications. Wire-bottom cages are always of the suspended type (Fig. 2.4). Solid-bottom cages may also be suspended, as shown in Fig. 2.5.

Animal Caging Systems 19

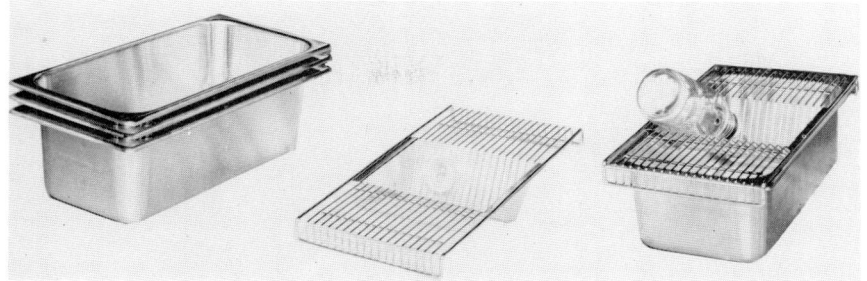

Fig. 2-3. Stainless steel solid-bottom mouse cages. (Photo courtesy of Hoeltge Bros. Inc.)

Fig. 2-4. Rolling rack of suspended wire-bottom cages. (Photo courtesy of Hoeltge Bros. Inc.)

20 *Materials and Methods for Breeding and Housing Laboratory Mice*

Fig. 2-5. Solid-bottom suspended plastic cages on rolling racks.

Fig. 2-6. Rolling rack with adjustable shelves displaying three common mouse cage types—polypropylene, polycarbonate, and stainless steel.

Once the cage type has been selected, the type of rack best suited can then be determined. Figure 2.6 is a rack showing several types of cages—polypropylene, polycarbonate, and stainless steel. It is of utmost importance to select cages and racks free of sharp edges, cracks, and corners that are difficult to clean. Also, the rack must be able to withstand repeated disinfecting and/or sterilization procedures. The rack shown (Fig. 2.6) has adjustable shelves, allowing for versatility in an animal colony.

Presently, glass bottles of various sizes are the most common watering system in use. The bottles may be handled separately in a special bottle-washing machine (Fig. 2.7a) or passed through a combination cage-and-bottle

Fig. 2-7 (a) Single-purpose bottle-washing machine. (b) Tunnel-type combination cage-and-bottle washer. (c) Cabinet-type combination cage-and-bottle washer.

22   *Materials and Methods for Breeding and Housing Laboratory Mice*

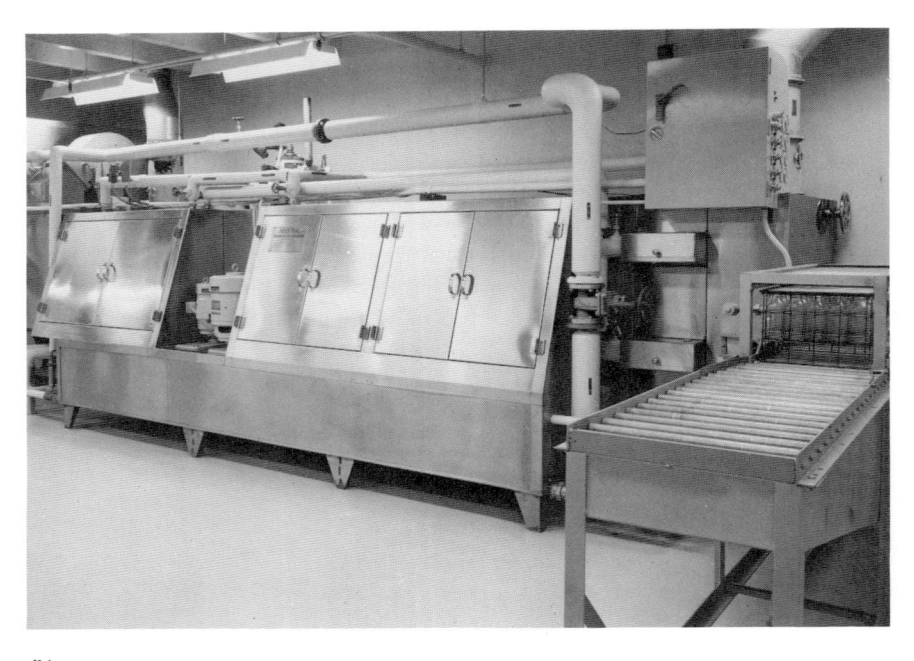

(b)

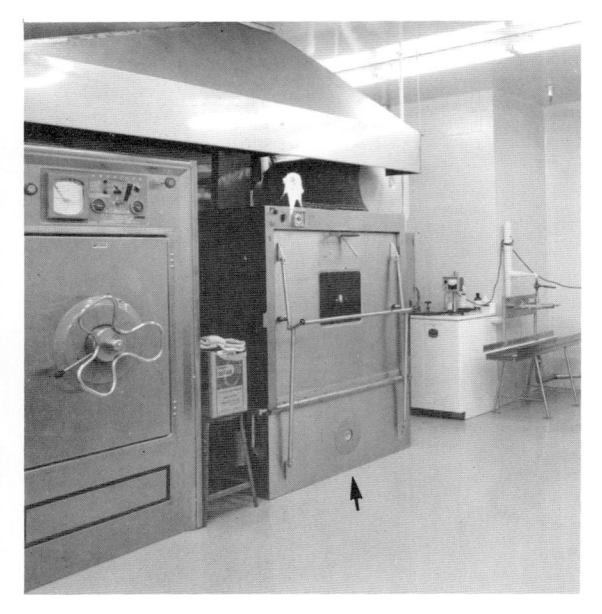

Fig. 2-7 (continued)

(c)

washer (Fig. 2.7b). If a combination unit is used, the bottles are washed first; then an adjustable hold-down bar is positioned prior to running the cages through a tunnel-type washer. In addition to the tunnel-type washers, single-batch washers are also available (Fig. 2.7c). Bottles are usually washed in stainless steel racks holding 25 bottles. A bottle-filling station (Fig. 2.8) may be located at the end of the machine. The supply of water may be plain tap water or chlorinated to a level of 14 to 16 ppm by a manual or automatic chlorination procedure (Appendix).

If not handled properly, bottle stoppers and sipper tubes can be a common source of *Pseudomonas* contamination. The inside of the sipper tube is the most difficult to clean. Sipper-tube washers are available commercially (Appendix) although a common method is to soak stoppers and tubes in a hot concentrated solution of detergent for about 30 min. Frequent agitation helps to dislodge particles of food or other solid matter that may collect inside the tube. The second step is to boil the stoppers and tubes in a pass-through steam-heated vat (Fig. 2.9) for 30 min.

The baskets of stoppers and tubes are then soaked in chlorinated water at a concentration of about 1200 to 1500 ppm for at least 30 min more. By repeating this process each time bottles are washed and refilled (usually three

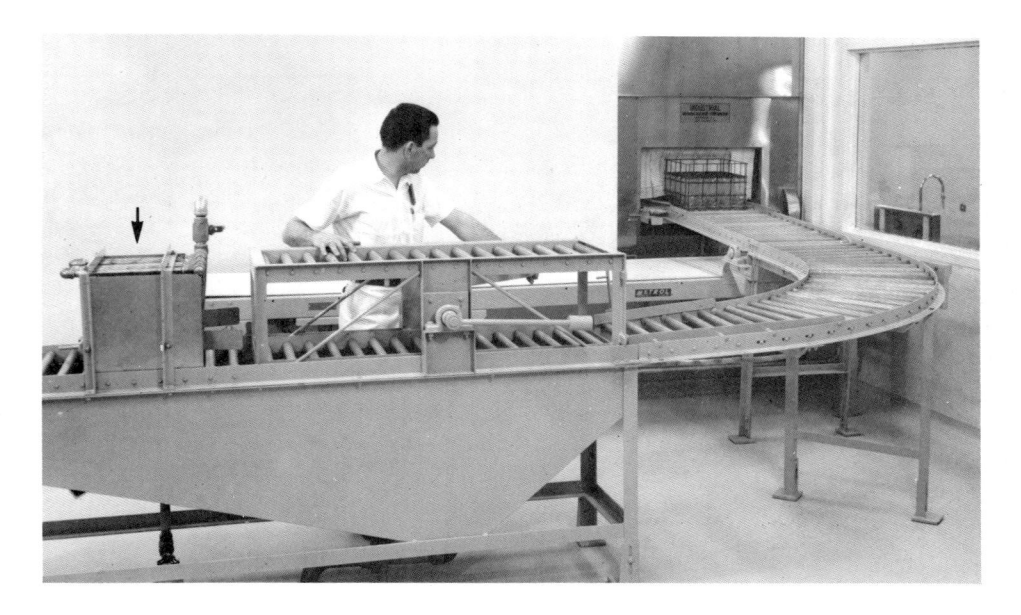

Fig. 2-8. A 25-bottle filling station with manual bottle-rack turner.

Fig. 2-9. Pass-through steam-heated stopper-boiler vat. Also shown are pass-through cage-rack washer, bottle-filling station, and bedding dispenser.

times per week), the growth of microbes can be controlled. A more reliable method is to autoclave the stoppers and tubes after washing.

Bedding

With the exception of the wire-bottom type, all cages discussed so far require one or another of the many kinds of bedding material presently available (Appendix). The most widely used materials include wood shavings, wood particulates, and ground corn cobs, although other products—such as sawdust, peatmoss, excelsior, shredded paper, treated cellulose fibers, and inert synthetics—have also been used and are available. The expanded wire-bottom cage also requires some type of absorbent material in the refuse pans under the cages. Thus, the final selection of bedding material to be used depends upon (1) the type of cage to be used, (2) the type of mice the cage will hold (breeder or experimental), (3) specific properties of the bedding—such as absorptive capacity, nonedibility, freedom from dust, and so forth—and (4) method of dispensing into cage and disposing of soiled bedding.

Inasmuch as the type of cage selected and its use will depend upon research requirements and personal selection, the important specific properties of the bedding itself must be considered. Bedding is used primarily for absorption; it must be capable of absorbing large amounts of urine and any water lost from the water bottle. Therefore, the bedding must have a relatively low moisture content, a feature that can create special problems. Very dry bedding is usually very dusty, and dusty bedding may be an antigenic stimulus and very irritating to both the mice and the caretakers. Very dry bedding also causes severe dehydration and death to newborn mice.

Bedding, which may be routinely used without being sterilized, should be as free as possible from contaminating microorganisms. Bedding used for germfree, defined-flora, or specific-pathogen-free mice must be capable of withstanding sterilization. Several common methods of bedding sterilization are used. Steam sterilization in a high vacuum sterilizer is the most effective method for wood shavings or ground corn cobs. With modern sterilizers, drawing of the vacuum (12 to 18 mm Hg) provides better steam penetration of the load. A second, less acceptable method is the use of ethylene oxide gas. The main problem associated with this type of sterilization is the ethylene glycol by-products remaining in the bedding, which may cause irritation and death owing to hemorrhagic diathesis (see Chapter 3).

Bedding Disposal

Several common methods for removing soiled bedding are employed today, and the selection is generally based primarily upon the total number of cages involved and the type of bedding selected. The simplest method is to dump or scrape the soiled bedding into large drums lined with strong plastic liners. The bags are then hauled away for disposal. A vacuum system, either in the room or at the cage washer, may be used to remove soiled bedding directly to an incinerator or removable dumpster. This system works well with the ground-cob type of bedding, and satisfactory results may be obtained with wood shavings as well.

Another method being used is the grinding and flushing of bedding material into the animal facility drainage system. Again, the system may be located either at the cage-washing machine or in the animal room in the form of automatic flushing valves, cascading water under each row of suspended wire-bottom cages. In barrier operations, two major problems are directly related to the latter method of refuse disposal. The first is maintaining a stable humidity in the room, and the second is controlling the growth of microorganisms and fungi resulting from the increased moisture.

General guidelines for proper sanitation and maintenance have been established. Included within these guidelines are facility as well as individual cage recommendations. The *Guide for Laboratory Animal Facilities and Care* (Public Health Service Publication No. 1024) recommends, under Sanitation Practices:

(1) Cleanliness
    (a) The animal facility should be kept clean. This means that a regular schedule of sanitary maintenance is necessary.

    (b) Animal rooms, corridors, storage areas, and other parts of the animal facility should be washed, scrubbed, vacuumed, mopped, or swept, using appropriate detergents and disinfectants, as often as necessary to keep them free of dirt, debris, and harmful contamination. A continuing objective should be to keep these areas neat and uncluttered.

(c) If litter or bedding is used in animal cages or pens, it should be changed as often as necessary to keep the animals dry and clean, and to minimize offensive odors. For simple maintenance of small rodents such as rats, mice, or hamsters, 1 to 3 such changes per week ordinarily should suffice. For larger species such as dogs, cats, and simian primates, daily changing of cage or pen litter may be necessary.

(d) Cages or pens from which animal waste is removed by hosing or flushing should be cleaned daily or oftener. This system may require removal of the animals during servicing in order to keep them dry.

(e) Animal cages, racks, and accessory equipment such as feeders and water bottles, should be washed and sanitized as often as necessary to keep them physically clean and free of harmful microorganisms. Ordinarily, this can be achieved by washing the cages and accessories once or twice weekly, and the racks every other week. In addition, cages should always be sanitized before new animals are placed in them. It is good practice to have extra cages available at all times to permit maintenance of a systematic cage-washing schedule. To reiterate a point . . . the washing or rinsing, or both, should be conducted at a temperature of 180 deg. F or higher to assure destruction of all pathogenic organisms except spore formers. If this temperature cannot be attained, washing of equipment should be followed by appropriate disinfection, using an effective disinfectant.

(f) Waste containers and implements should be maintained in a sanitary condition. It is good practice to wash each waste can every time it is emptied. As with animal cages, the minimum wash or rinse temperature, or both, should be 180 deg. F. Similarly, cleaning implements such as scrapers, shovels, and mops should be sanitized regularly.

This guide further recommends 0.1 to 0.7 sq ft per mouse in a small group cage.

We use two to three cage changes per week, depending on the number of mice per cage, and three water bottle changes (using chlorinated water, 16 ppm), which is quite satisfactory. It is important that each animal facility have an empty room that can be decontaminated and used in rotation, so that all rooms can be sanitized and decontaminated on a rotating basis.

BREEDING

In mature mice, the estrous cycle occurs every 4 to 5 days. There may be a postpartum estrus within the first 24 hr. Most strains of mice are mature by about 8 weeks of age. In some strains the males may mature earlier. The period of gestation is 19 to 21 days, except in breedings during postpartum estrus, in which case it may be extended. Different breeding methods include breeding pairs, breeding trios (one male to two females), and harems (one male to four or more females). The single-pair mating system is probably the best, but there may be cases when space utilization and production needs make trio breeding desirable. Harem breeding is not practical from a disease control and sanitation point of view unless the pregnant females can be removed to other caging for delivery of their young.

Male mice ordinarily do not interfere with the young and are usually left in the cage with the female throughout her entire breeding life. This practice may tend to increase the total number of offspring owing to breeding that occurs in the postpartum estrus. It is sometimes referred to as "forced breeding." Choices between inbreeding, random breeding, or hybrid breeding are intimately related to the experimental subject and experimental design.

Systems

Inbreeding systems have been used for many years to increase the uniformity of animals and therefore hopefully to reduce the experimental variation. There are many experiments involving tissue transplantation and radiation protection where inbred animals that will readily accept organ and cellular grafts are quite necessary. In addition, of course, many genetic experiments require inbred animals. Inbreeding systems include brother–sister matings generation after generation and parent–offspring matings generation after generation. Mice of a given strain are considered inbred after 20 or more consecutive generations of brother–sister matings, or 20 or more consecutive generations of parent–offspring matings if the offspring were mated to the younger parents (STAATS, 1964). Crosses between two different inbred strains are F_1 hybrid mice. They are often considered much more vigorous, and many F_1 hybrids will accept grafts from both parent strains. They are very useful in genetic experiments inasmuch as they are uniformly heterozygous, whereas inbred strains are uniformly homozygous.

Many mouse colonies that are not actively being inbred are considered random bred. However, this appellation is not strictly accurate, for true random breeding requires that tables of random sampling or other methods be used to insure random mating is actually done.

For additional information on breeding methods and recordkeeping systems, an excellent reference is *Biology of the Laboratory Mouse*, especially Chapters 2 and 3 (see reference section). In Chapter 5 of this book (Technological Advances), reference is made to some specific computerized breeding programs that use both inbreeding and hybridization systems.

HOUSING

Conventional

Conventional mice comprise a large group of animals that can be housed in a number of different ways. Some colonies of essentially conventional animals are housed in very high-quality facilities with elaborate corridor and air control systems, similar to those described for SPF mice. This situation does allow the possibility of eventually converting the conventional facility into an SPF facility by replacing all animals with hysterectomy-derived animals.

Most conventional colonies, however, usually consist of single rooms on a common corridor, often housing several different kinds of animals within the same facility. The air conditioning, if present, usually does not provide 100% fresh air makeup, which is desirable for laboratory animals, but often is only 10 to 20% fresh air makeup. Despite the handicaps of the usual conventional mouse facility, a mouse acceptable for many experimental procedures can be cared for in such facilities if some ingenuity is used. We have housed SPF animals in a conventional facility, using only window air conditioners and steam heat. Using filter-cap caging and strict sanitation procedures, we were able to keep the mice in this room free of all pathogens for an 18-month period. Many experiments could be accomplished, of course, in much less time.

A subject of considerable importance to the investigator and to the management of animal facilities is the location of the animals in relation to the investigator. There are generally two divergent opinions: one is that small independent facilities should be located near the investigators' laboratory; the other is that centralized facilities should be placed at some common point, or midpoint, for use by everyone. There are good arguments for both systems. If a new research facility is to be built, the methods can be combined, and a centralized animal facility can have laboratories placed around it in such a way that it can be centrally operated and the investigator can have animal space immediately adjacent to his laboratory. However, what usually happens is that the animal space is added long after the laboratory space; or a new investigator arrives, and laboratory and animal space must be found for him at the same time.

Advantages of having small adjacent animal facilities are limited to convenience to the investigator and his closer observation of the animals during an experiment. However, a central facility can offer better service at the same or reduced cost because of centralized equipment and reduced man hours. It is more efficient to have one individual responsible for all the animals. There are exceptions, of course, and each situation should be evaluated on the basis of economics and practical considerations.

Germfree, Defined-Flora, and Specific-Pathogen-Free

Many technological advances have been made in germfree housing in recent years. Inasmuch as most of the principles are now well understood, the information is readily available to anyone desiring to use it. We have attempted to diagram the basic principles (Fig. 2.10). In general, the ambient

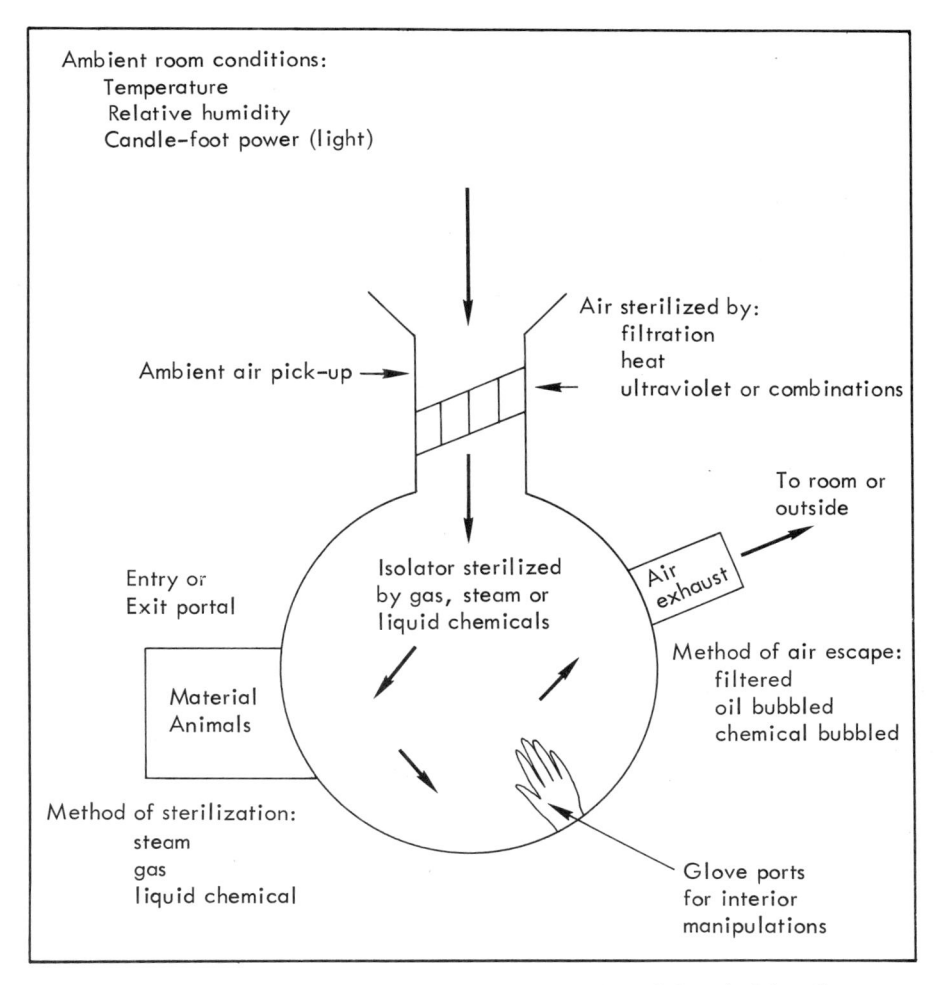

Fig. 2-10. Diagrammatic representation of the principles of germfree housing.

room conditions must be controlled with respect to temperature, humidity, and light in order to control the conditions within the isolator. Exceptions exist, of course, as with isolators with inside lighting systems. The ambient air is picked up and sterilized by a single method, or by combinations of methods. The ambient air is force-filtered through a series of fiberglass insulation layers (Fig. 2.11). The sterilized ambient air then enters the previously sterilized isolator chamber. The materials used to presterilize the chamber include gas, steam, and liquid chemicals. The air must be evacuated from the sterile chamber without risk of backflow contamination. Several methods for air exhaustion are used, including filtrations, bubbling through oil (Fig. 2.12), bubbling through chemicals, and backflow-prevention devices.

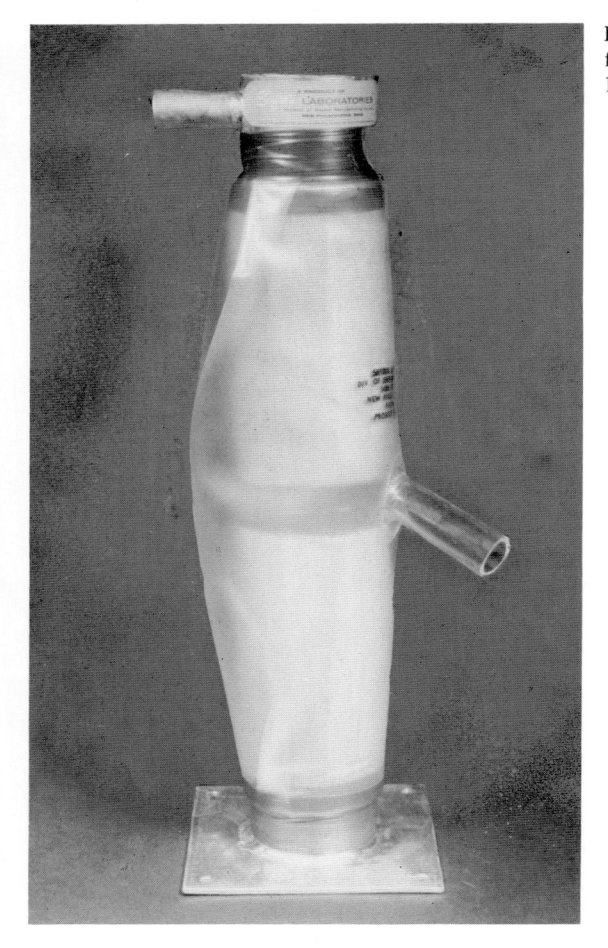

Fig. 2-11. Incoming air filter for germ-free isolator. (Photo courtesy of Snyder Manufacturing Co.)

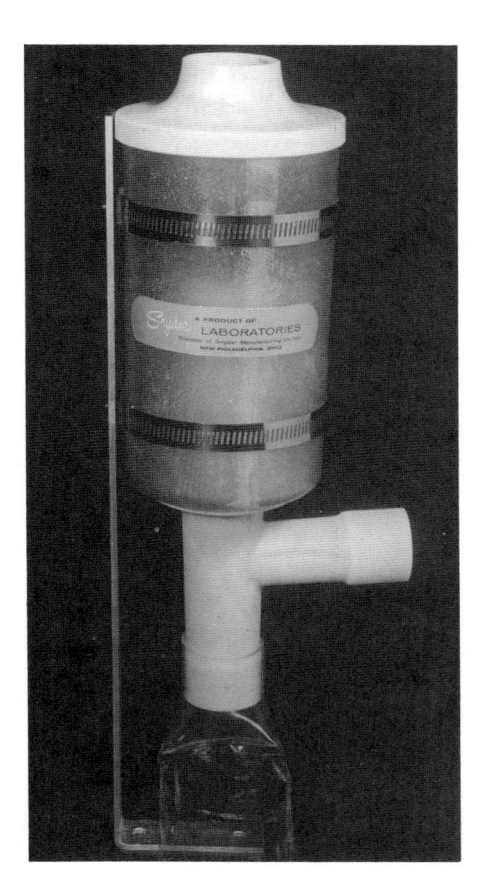

Fig. 2-12. Oil trap for germfree isolator exhausted air. (Photo courtesy of Snyder Manufacturing Co.)

Despite the source of germfree or isolator-housed gnotobiotic mice, or which of the many different sterile isolator housing units is used, the housing techniques have certain necessary factors in common. For example, all systems must have a sterile air supply, a means of exhausting air from the unit without backflow contamination, and methods of introducing mice and materials into the unit without concurrent contamination.

Three basic types of isolators and three basic types of entry systems are discussed. The reader should bear in mind that these entry systems and sterile isolator units can be—and usually are—used in different combinations. For example, a single isolator may have a steam sterilization entry lock and a liquid chemical and/or gas sterilization entry port, or just a gas sterilization entry port (Fig. 2.13). Other units may have only a single entry system. In general, the type of system or systems used depends on the nature of the

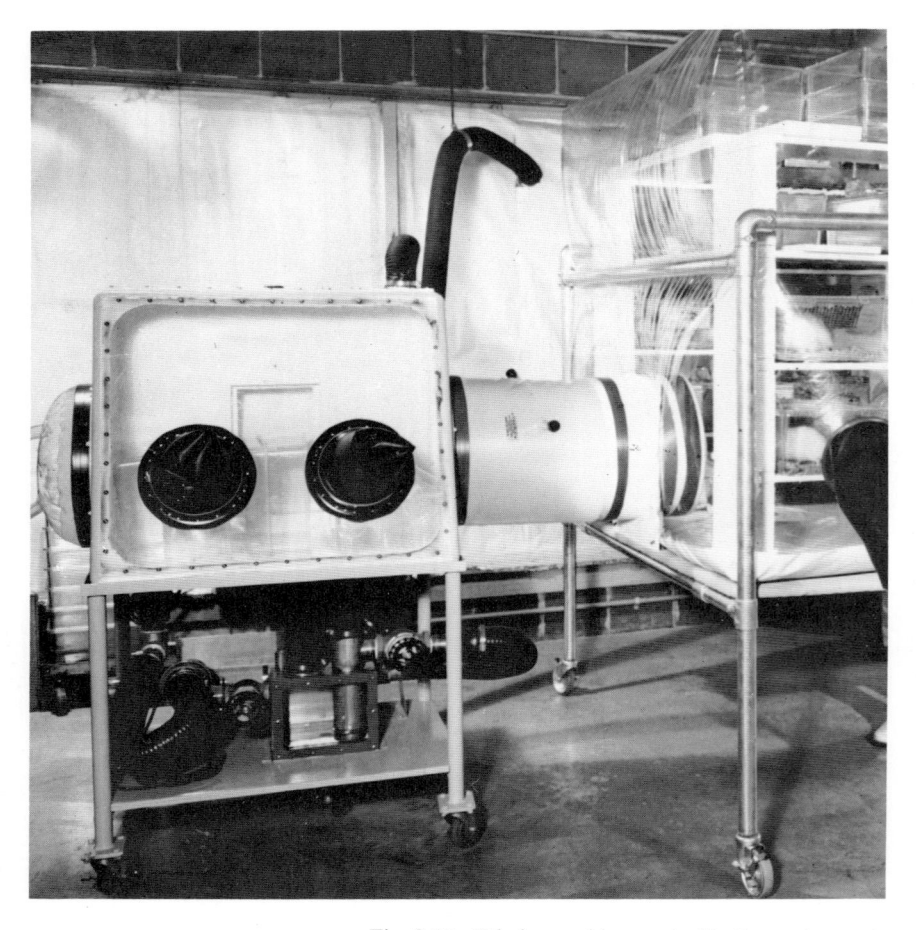

Fig. 2-13. Ethylene oxide gas sterilization entry port.

experimental work, personal preference of the investigator, and the economics of the particular situation. Table 2.1 summarizes the basic methods to be discussed.

Almost every combination has been used at one time or another except the steam autoclave entry portal attached to a plastic isolator, an obviously unworkable combination. The early workers in the field used primarily steel or stainless steel as construction material.

Some of the most interesting early germfree work was done by MIYAKAWA (1959), who used a very large steel tank with remote control devices for manipulating animals and materials inside (Fig. 2.14). A standard stainless

TABLE 2.1

Basic Isolators and Entry Systems

Three basic isolator types	Sterilization entries
Metal	Steam
Steel	Presterilized drums
Stainless steel	Autoclave entry portal
Hard plastic	Liquid chemical
	Dip vat
	Spray or mist in transfer sleeves
Flexible plastic	Gas
	Autoclave entry portal
	Inside isolator

Fig. 2-14. Complex steel isolator with remote controls, used by MIYAKAWA (1959). (Photo courtesy of Dr. M. Miyakawa, University of Nagoya, Japan.)

Fig. 2-15. Reyniers' stainless steel isolator with steam sterilization entry port. (Photo courtesy of Dr. R. Zinn, National Institute Health, Bethesda, Md.)

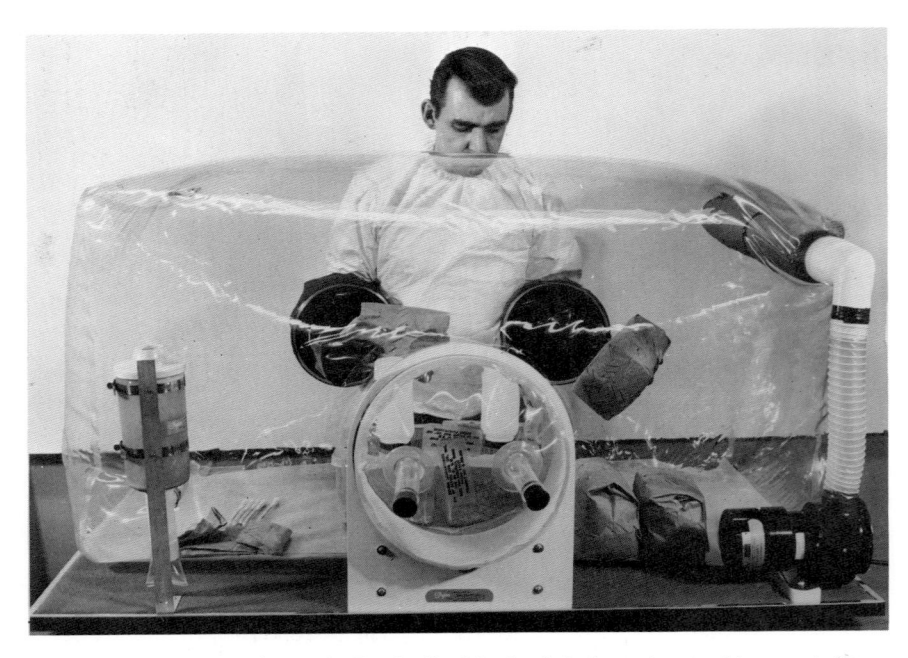

Fig. 2-16. Trexler flexible plastic isolator showing blower, air filter, entry portal, and exhaust trap. (Photo courtesy of Snyder Manufacturing Co.)

steel unit is the Reyniers' germfree isolator (Fig. 2.15), constructed of 12-gauge metal, viewing ports, steam sterilizer supply lock, air filters, steam supply, and drain and electricity inlet. The Luckey-Blickman isolator is also a stainless steel isolator having both steam and liquid chemical entry ports (LUCKEY, 1963, p. 116).

A second basic style isolator is the Trexler flexible plastic isolator (Fig. 2.16), which is extremely versatile and can be built relatively inexpensively in various shapes or sizes, depending on the experimental design and on animal production needs. It is particularly useful for a production situation. In our laboratory, we have used a large plastic isolator having a center suit with its own air supply (Fig. 2.17).

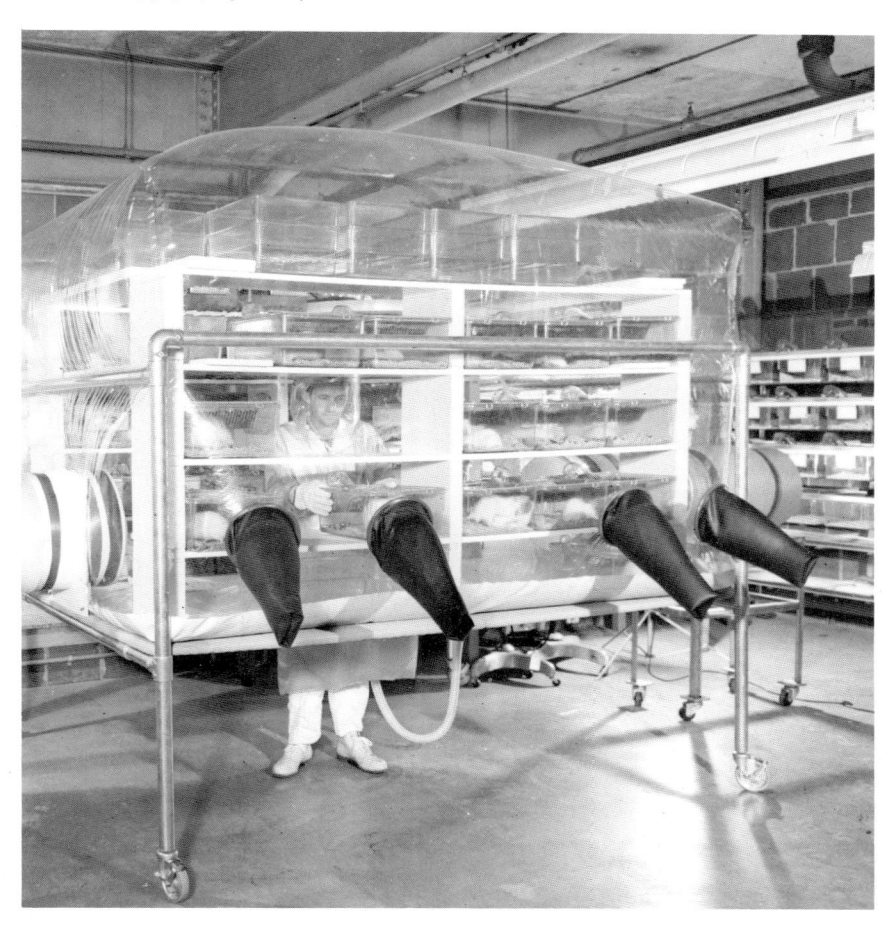

Fig. 2-17. Large tent plastic isolator with central service unit.

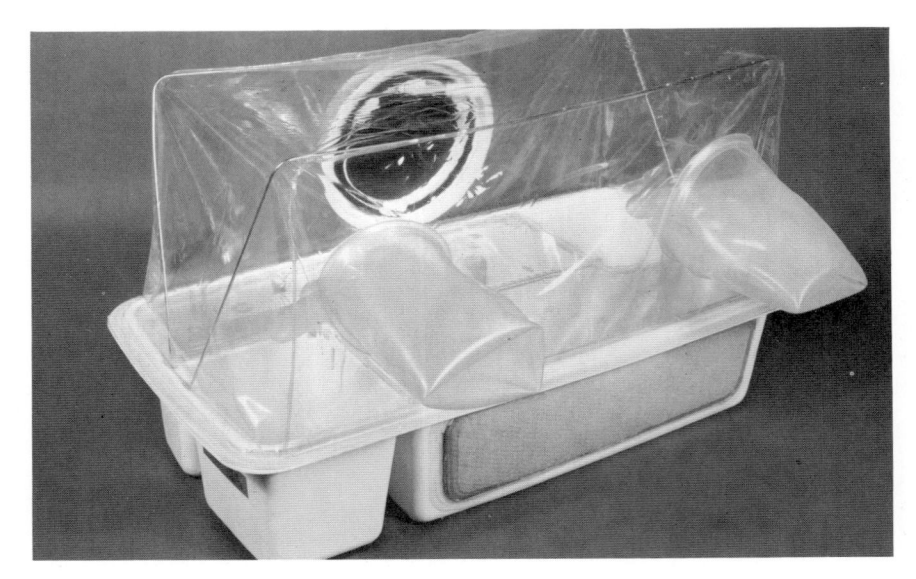

Fig. 2-18. Isolab isolator—small, compact, and free of mechanical air-handling system. (Photo courtesy of Lab-Cages Inc.)

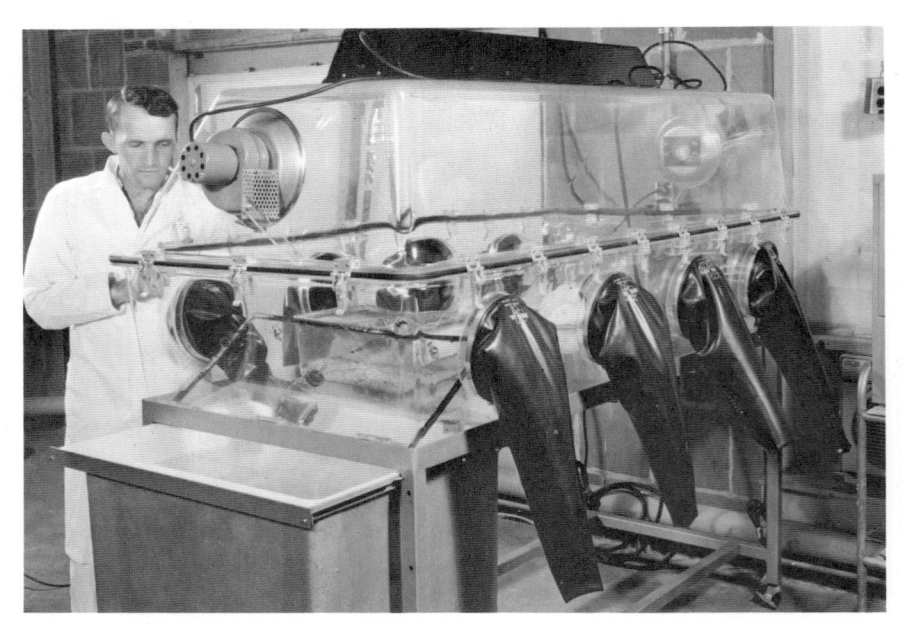

Fig. 2-19. Rigid Plexiglas isolator with liquid-disinfecting entry system.

Fig. 2-20. Lightweight metal chambers ready for sterilization; developed by PILGRIM and THOMPSON (1963). (Photo courtesy of Dr. Ira Pilgrim, University of Utah, College of Medicine, Salt Lake City, Utah.)

The Isolab (Appendix) should also be mentioned (Fig. 2.18). It is a small disposable isolator, without a blower system, useful in many ways when small temporary sterile isolators are needed. Made of flexible plastic, it may house 20 to 30 mice.

A third basic type is the rigid plastic isolator (Fig. 2.19), as described, for instance, by PHILLIPS *et al.* in 1960. These rigid plastic isolators lend themselves very well to a liquid chemical dip tank entry method. In addition to their use as sterile isolators, they can, unlike the soft plastic isolators, be changed to a negative pressure system and be used for containment experiments.

All the types of isolators just described are available commercially; however, experience has shown that successful germfree work can be done, and money saved, in ingenious and imaginative ways. A method developed by PILGRIM and THOMPSON (1963) at the University of Utah involves lightweight metal chambers completely filled with all necessary food and canned water, sealed with Mylar film and steam sterilized (Fig. 2.20). In this case, only the mice need to be transferred from one isolator to another, and no materials

Fig. 2-21. Preparation of Pilgrim lightweight metal isolator for steam sterilization. (Photo courtesy of Dr. Ira Pilgrim, University of Utah, College of Medicine, Salt Lake City, Utah.)

need to be introduced or taken out during the experiment. Figure 2.21 shows a free-flow design of a similar isolator system by PILGRIM (1969). Both styles of Pilgrim isolators can be set up with battery-powered systems.

Almost any caging system can be used, provided it can be passed in and out of the isolator conveniently. In actual practice, a plastic cage is often used because animals can be seen much better. The investigator's personal preference must be considered when selecting caging.

Of major importance in considering a barrier facility designed to house DF or SPF mice are the specific functions that the facility will be expected to fulfill. To provide a comprehensive working plan for this design, one must consider carefully the details of all the functions that will eventually be carried out within the barrier. This plan must include the number of animals required, equipment needed, cost, and above all, the ability to maintain the mice with the exact preselected flora. Compared with SPF mice, DF mice require more absolute environmental control and more specialized procedures within the facility. In the formulation of such a plan, it is necessary to include the thinking of research personnel who will use the facility, administrative personnel, and the professional personnel responsible for the daily operational procedures.

Once the functions or requirements of the animal research facility are established, the architect and engineers are included in subsequent planning. During this stage of planning, specific applications of the facility should be discussed. For example, will the facility be used for production of several species of rodents, for long-term holding of aging animals, for the production of biologicals, for specialized research involving infectious viruses or bacteria, for lethal and supralethal irradiation studies, or any combination of these?

Several important points to keep in mind concerning the overall design are: (1) the entire facility and equipment should be kept on the same level, if possible; (2) internal construction material must be durable and easily cleaned; (3) the facility must be rodent- and vermin-proof; (4) flexibility and

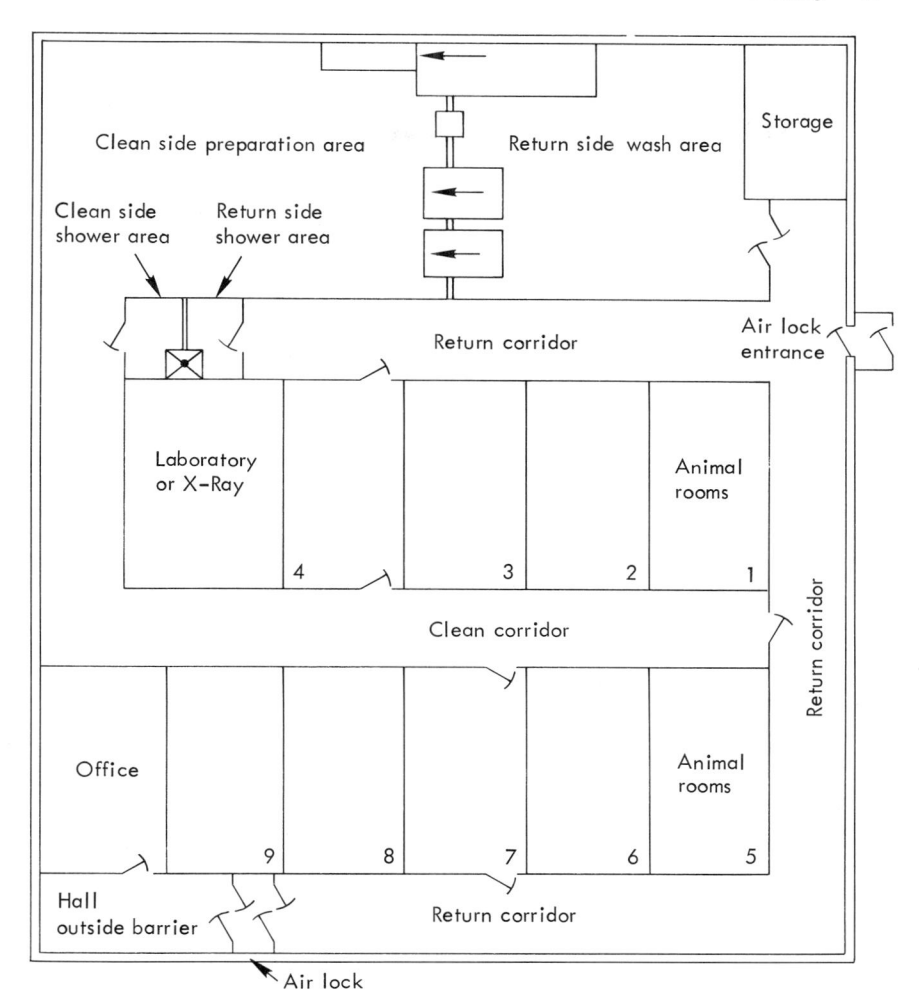

Fig. 2-22. Diagrammatic barrier floor plan employing single central clean-corridor system.

complete control of the total air-handling system, including an emergency source, are mandatory.

Barrier isolation facility designs usually encompass one of the following general layouts: (1) a single central clean corridor, with surrounding return corridors (Fig. 2.22); or (2) a two-corridor system, one clean and one return, primarily used in long narrow buildings (Fig. 2.23). In a recent publication, JONAS (1965) described six different types of laboratory animal facility designs,

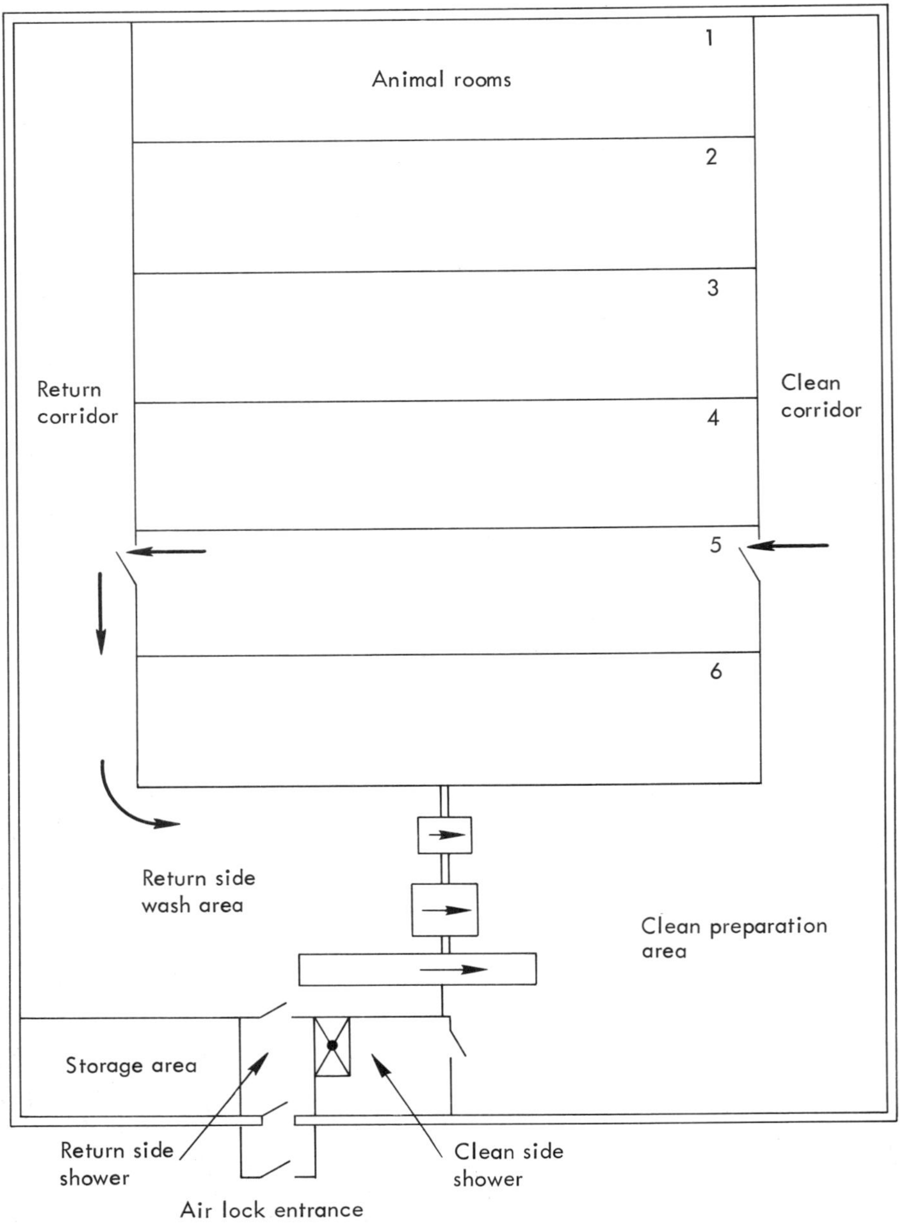

Fig. 2-23. Diagrammatic barrier floor plan utilizing one clean and one return corridor.

and two recent papers by SIMMONS *et al.* (1967) and BRICK *et al.* (1969) describe facilities utilizing the double and single central corridor ideas.

Also of major concern in such designs is the relationship of certain equipment (cage washer, sterilizer, etc.) to the animal rooms, of research laboratories to the animal rooms, and of the general type of work to be carried out within the facility, that is, long-term aging studies, acute studies (survival or nonsurvival), or production.

In a barrier facility, the cage-and-bottle washer, sterilizer, rack washer, and so forth, are usually of the pass-through type. All of this equipment should be on the same level as the rest of the barrier, although sterilizers located on the floor above the facility have been tried. As shown in Figs. 2.22 and 2.23, this type of equipment opens into clean preparation areas. All doors should be electrically integrated so that only one door can be opened at a time, thus reducing the possibility of air flow, unsterilized material, and so on, passing from the return side to the clean side. The model of cage washer to be used will be based primarily on the available space, type, and number of cages to be washed and, of course, the cost of the machine. For example, if space permits, a tunnel-type combination cage-and-bottle washer may be used. This type of machine can be purchased with a complete outside covering of stainless steel lined with 1-inch-thick polyurethane insulation (Appendix), which effectively reduces the surface temperature from 200 to 80F as well as indirectly eliminating the need for extra air-conditioning units to handle the cage-and-bottle washing area.

The sterilizer, again depending on available space, may vary considerably in size. Recent models are available with high-vacuum (12 mm Hg absolute) cycles that allow for more efficient sterilization of feed, bedding, and so on (Fig. 2.24). Also available are combination steam and gas (ethylene oxide)

Fig. 2-24. Large pass-through steam sterilizer with a load of bedding, as seen from the return side of barrier.

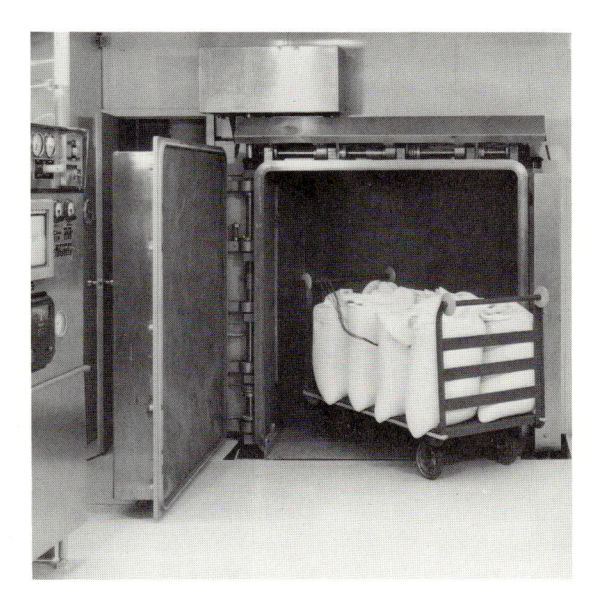

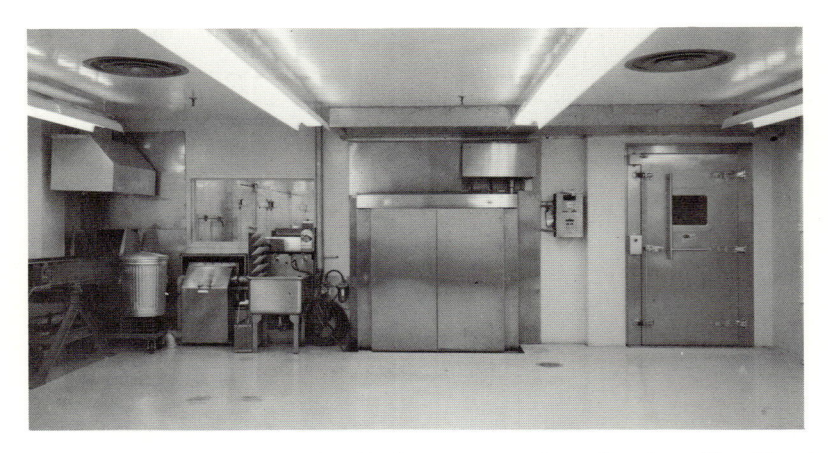

Fig. 2-25. Equipment arrangement for clean side of barrier. Left to right, opening from cage-and-bottle washer, pass-through stopper boiler vat, sterilizer, and cage-rack washer.

sterilizers that allow considerable versatility (Appendix). A pass-through cage rack washer may be desirable (Fig. 2.25), and many models are available (Appendix). Final selection will usually be made on the basis of space, desired use, and cost.

A method of disposing of soiled bedding must be devised, and systems range from dumping bedding into steel drums lined with large plastic bags to automatic systems (pneumatic or auger) that carry the bedding directly to an incinerator, dumpster, or large drain.

Construction material for isolation facilities varies widely. The one thing that is generally agreed upon is the avoidance of any wooden structures. Frequent cleaning and disinfecting procedures require material such as smooth concrete or block walls covered with epoxy, polyurethane, or dry wall treated with a waterproof sealant. Of utmost importance is a smooth, durable finish that, if possible, can be integrated into the floor surface. The floors may be constructed of epoxy-covered cement, terrazzo, acid-proof brick, or smooth-finish cement covered with polyurethane applied the same as on the walls and ceiling. If floor drains are used they must be installed with smooth, tight-fitting coverplates to reduce the possibility of contamination from the drain system. Certain types of floor coverings, when wet, are very slippery and dangerous. Most paints, such as epoxy or polyurethane, should have sand added to the final coat.

Special construction details, unique for barrier facilities, include vapor-tight, flush-mounted ceiling lights that can be changed from an overhead plenum. However, this process is very expensive, and vapor-proof lights can be suspended much less expensively. The vapor-proof light allows for complete room decontamination by spraying with a liquid or gas disinfectant. Also, soft, nonglare fluorescent lighting of sufficient foot-candle power should be used.

The electrical outlets within the rooms should also be flush mounted and vapor tight for ease of complete room cleaning. To prevent entry of contaminated air or vermin through the electrical conduit system, a nonhardening caulking compound should be placed in the conduit after the wires have been pulled through.

Other features that one may incorporate in the design of the animal rooms include sinks within each room. Although sinks may be an additional source of contamination, they are useful during routine cleanup procedures, especially if there are no floor drains. Also to be considered are (1) flush-mounted intercommunication speakers for emergency purposes as well as daily operational requirements, (2) a central vacuum system for removing soiled bedding within the room, and (3) positive air transfer hood or a transfer port connected to isolators in adjoining laboratories for work with infectious agents.

Construction features incorporated within the barrier, but preferably on the return side, include manometers to monitor individual room pressure changes or specific areas within the barrier, that is, the clean-side preparation area in direct relation to the return-side area. A more complex system may be installed to monitor all pressure differentials within the barrier directly to an alarm and control panel located in the supervisor's office. Individual room thermostats (for specific room temperature control) should be located outside each animal room, usually in the return corridor. Automatic lighting-control boxes can be located outside of each individual room, or they may be grouped together at a convenient location on the return side. The more sophisticated the control system, the more expensive it is to install and the more complicated to maintain. However, a good control system may save dollars in labor costs, including air-conditioning, temperature, and humidity control of incoming air supply.

The design of environmental controls throughout a barrier system is a most important consideration. In a situation where future research requirements may change, it is equally essential to design an air-conditioning system that is flexible. It is important to the research animal facility in order to meet whatever environmental demands may arise; an environment that is precisely controlled will contribute significantly to the overall success of the research program. The type of air-conditioning system selected and the degree of environmental control required for each facility vary considerably; therefore, we have included a description covering the major aspects of a complete system. Again, a close working relationship between the scientific

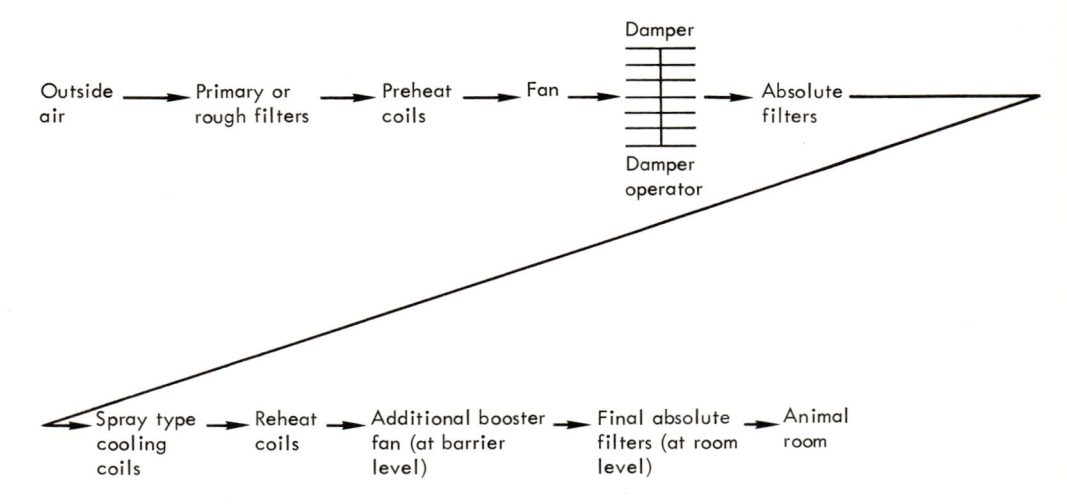

```
Outside ──▶ Primary or    ──▶ Preheat ──▶ Fan ──▶ [Damper] ──▶ Absolute ──▶
air         rough filters     coils                            filters
                                                  Damper
                                                  operator
```

```
──▶ Spray type  ──▶ Reheat ──▶ Additional booster ──▶ Final absolute   ──▶ Animal
    cooling         coils      fan (at barrier         filters (at room     room
    coils                      level)                  level)
```

Fig. 2-26. Diagrammatic arrangement of air-conditioning and air-handling system for any type colony.

staff and the mechanical system designer or architect designing the facility should provide an air-conditioning and environmental control system adequate for the particular facility.

The system to be described provides 100% outside air, which is prefiltered, preheated or cooled, reheated as regulated by the room conditions, absolute filtered, and discharged into the room. A general arrangement for this type of system is shown in Figure 2.26.

The rough or primary filter bank is made up by combining $24 \times 24 \times 8$-inch filter elements. These filters, at a 1000 CFM flow, have an efficiency rating of approximately 80% by weight. The incoming air, if necessary, is then heated to about 50F by single-row, nonfreeze, steam-type preheat coils. Vertically mounted, the coils are trapped and controlled to prevent condensate from freezing within the coil. During warm weather, eight-row multicircuit chilled-water cooling coils reduce the temperature of the incoming air from outside conditions to 50F dew point temperature. This cooling procedure is accomplished with chilled water supplied at 40F. The spray-type coils also increase the humidity of the incoming air, if necessary.

Manual adjustment of pneumatically operated dampers provided at the fan discharge area regulates pressures within the facility and compensates for the increased resistance acquired by the air filters as they become clogged with dirt. For example, by adjusting these dampers the cleanest area (Table 2.2) is maintained at a pressure that is more positive than any other area.

As a result of this type of setup, backflow that may occur will be toward an area with the same pressure or one of lower pressure (less clean area, such as a return corridor). The first bank of absolute high-pressure filters (99.9%) is located in the system just after the manual dampers. The filters are designed to remove any particles greater than 0.3 μ in size. Single-stage 440-v electric reheat coils, located in the duct supplying the room, are regulated automatically by sensing elements contained in the exhaust duct of each room. Room lights, mice, and personnel activity cause an increase in room temperature; therefore, the exhausted air temperature activates (controls) the reheat coils to maintain the desired preset room temperature. Recommended mouse room temperatures range from 70 to 75F, depending upon the major type of activity in the room. A second bank of high-efficiency particulate air filters is provided for maximum filtration. These filters are usually located directly over the animal rooms, and all ducts downstream from the final filters are constructed of aluminum to reduce the possibility of particulate matter getting into the room. Owing to the increased resistance from the filters located near the animal rooms, a booster fan may be incorporated into the system at a location just before the ductwork branches into the individual room air supply. A spare fan started automatically by failure of the first one insures a continuous air supply and positive pressure in this very critical area. Air being supplied to other areas (clean preparation, return side, etc.) of the barrier may come from separate ductwork, designed without the booster fan and second bank of absolute filters.

Rooms are supplied with a sufficient volume of air to provide a minimum of 12 to 15 changes per hour as required for odor control. Air may be introduced into the top of the room via a continuous slot diffuser running the entire length of the room or through several circular diffusers centrally located in each room. Of major concern is the resulting flow pattern that is designed to aspirate air from various cage levels without excessive draft

TABLE 2.2

Summary of Environmental Criteria

Area	Relative cleanliness	Relative pressure	Temperature (°F)	Humidity
Clean preparation	Ultra clean	+ + + +	71 ± 3	50% ± 5
Clean corridor	Ultra clean	+ + + +	71 ± 3	50% ± 5
Animal rooms	Ultra clean	+ + +	71 ± 3	50% ± 5
Return corridors	Clean	+	70 ± 3	No requirement
Return side wash	Clean	+	Comfort	No requirement

throughout the room. Therefore, with a change in cage-rack placement, one may expect a variable disruption to the air flow within the room.

Air may be exhausted from within the animal rooms by vents located in the ceiling or preferably at each end of the room near the floor. This exhausted air may be reused in less clean areas, for example, in the return-side wash area. In this way, all air supplied to the building is used for maximum benefit before being discharged to the atmosphere. Inasmuch as the ventilation rate in the wash area may be as high as 60 changes per hour, reusing exhausted air from the animal rooms is an effort to save money required for the additional refrigeration necessary for this area. Of course, there should be no recirculation of air from any infectious disease area, or from any area known to house contaminated mice. It is important to keep the exhaust fans located as far as possible from the incoming air supply.

Special exhaust systems may be required in certain situations, such as use of the cage-washing machine, radioisotope hoods, or during work with highly infectious material. In the latter case, the exhausted air may be required to go through an incineration process before being discharged to the outside.

Alarm systems are as necessary as any part of the air-handling system, because variation of a preselected environmental condition must be confirmed and corrected immediately. Therefore, the location of the visible alarm panel is of critical consideration. What part or parts of the entire system must be monitored to assure complete environmental coverage? The following discussion is based primarily on an existing alarm system that parallels the air-handling system outlined in this section.

The sensing unit for the temperature alarm is in the exhaust duct of each animal room. This unit is activated by a 3-deg drop or increase in room temperature. In the system described, a round-the-clock maintenance-and-utilities group receives the initial alarm, both visibly and audibly. They, in turn, call the facility supervisor who must decide what corrective measures to take (e.g., moving mice to another room, calling maintenance personnel, etc.). All alarm systems will be similar to this general format; therefore, the procedure will not be discussed again.

The sensing device for the humidity alarm is in an exhaust duct common to all the animal rooms. The sensing device is set to alarm at $\pm 5\%$ difference in the preselected room humidity.

Several alarms should be associated with the chlorinated water supply. The first alarm is electrical failure of the water pump. Two others are concerned with either high or low chlorine content. A flashing red light, visible from the water filling station within an isolation facility, is adequate warning to the animal caretakers not to use the water. Many other alarms can be incorporated, especially on the air-conditioning unit itself. A few additional alarms to consider are (1) failure of air flow (velocity-sensing device), (2) failure of fan motor (electrical-sensing device), (3) failure of water pump motor on spray coils, (4) process and chilled water pump failure, and (5) high or low steam pressure alarm.

Problems associated with a barrier facility are compounded by the fact that everything going into such an area must be sterilized. Obviously, because personnel entering this area cannot be "sterilized," they become the most common source of undesirable contamination. For a better understanding of this problem, a general day-to-day description of a typical barrier operation will follow. It is not intended to represent an ideal, but only to serve as a point of illustration.

A barrier may be designed with several concepts in mind. Normally, personnel enter a barrier through an entrance airlock and then into the shower area before entering the ultra-clean side. Street clothes are left on the return side of the shower. Following a modified surgical scrub, the men don sterilized coveralls, caps, face masks, and gloves on the clean side. Boots, disinfected in a germicidal dip tank while the men are showering, or previously sterilized shoes that remain inside the barrier, are usually worn.

Vitamin-fortified feed is sterilized in the manufacturer's bag or on multiple trays with a high-vacuum, steam, pass-through autoclave. Bedding must also be sterilized, usually in the bag. The laundry, prewashed and wrapped, must be sterilized before usage in the barrier. All material (cards, program lists, etc.) unable to withstand steam sterilization may enter the clean side of the barrier through an ethylene oxide gas sterilizer.

Personnel and material enter each room via the ultra-clean corridor. Dirty cages, bottles, racks, and so forth, are collected in the return corridor adjoining all animal rooms. Dirty cages and bottles are received in the return-side preparation area before passing through a cage and/or bottle washer into the clean side. Large rolling racks, carts, dollies, and the like, re-enter the clean side through a pass-through cage-rack washer. Bottle stoppers and sipper tubes enter the clean side through a stopper-boiler pass-through vat, or via a pass-through steam sterilizer.

Mice inside the barrier may be housed in various types of cages and bedding. Breeder cages holding just a pair or a pair plus a litter are changed twice per week. Experimental mice housed 10 per cage should be changed three times per week. Water bottles, if filled with water containing 14 to 16 ppm chlorine, should be changed three times per week to effectively control *Pseudomonas* sp. n. The animal rooms are cleaned daily and the floors are scrubbed weekly. The clean-side preparation area and clean corridors are scrubbed at least three times weekly. The wash room area and return corridors are scrubbed two to three times a week.

Various specialized equipment, such as an X-ray machine, may be included in the barrier to be used routinely for research involving SPF mice produced there. A laboratory may also be maintained within the barrier complex for procedures involving the SPF mice. Mice should be handled in both of these areas so as to prevent contamination, thereby allowing them to be returned to experimental holding rooms within the barrier.

CHAPTER 2 REFERENCES

Brick, J. O., R. F. Newell, and D. G. Doherty. "A Barrier System for a Breeding and Experimental Rodent Colony: Description and Operation." *Lab. Animal Care* **19**: 92–97, 1969.

Green, Earl L., ed. *Biology of the Laboratory Mouse*, 2nd ed. (New York: McGraw-Hill Book Company, 1966), 706 pp.

Jonas, A. M. "Laboratory Animal Facilities." *J. Am. Vet. Med. Assoc.* **146**: 600–606, 1965.

Luckey, Thomas D. *Germfree Life and Gnotobiology* (New York: Academic Press, 1963), 512 pp.

Miyakawa, Masasumi. "Report on Germfree Research at the Department of Pathology, University of Nagoya, Japan, and Some Observations on Wound Healing, Transplantation and Foreign Body Inflammation in the Germfree Guinea Pig," in *Recent Progress in Microbiology*, Symposia held at VII International Congress for Microbiology, Stockholm, 1958, ed. by G. Tunevall (Springfield, Illinois: Charles C Thomas, Publisher, 1959), pp. 299–313.

Phillips, A. W., F. A. Rupp, J. E. Smith, and H. R. Newcomb. "A Plexiglas Isolator for Germfree Animal Research," in *Proceedings of the Second Symposium on Gnotobiotic Technology* (Notre Dame, Indiana: University of Notre Dame Press, 1960), pp. 49–54.

Pilgrim, H. Ira. "A Lightweight Autoclavable Germfree Mouse Isolator." In press, 1969.

Pilgrim, H. Ira, and David B. Thompson. "An Inexpensive, Autoclavable Germfree Mouse Isolator." *Lab. Animal Care* **13**: 602–608, 1963.

Simmons, M. L., L. P. Wynns, and E. E. Choat. "A Facility Design for Production of Pathogen-Free, Inbred Mice." *ASHRAE* **9**: 27–31, 1967.

Staats, J. "Standardized Nomenclature for Inbred Strains of Mice." *Cancer Res.* **24**: 147–168, 1964.

Animal Health
THREE

INTRODUCTION

The problem of animal health in a mouse colony has, of course, many ramifications. This chapter is devoted to major problem areas—noninfectious diseases, infectious diseases, disease prevention and control, nutrition and nutritional deficiency, pest control, and, most important, animal testing and monitoring. In certain clear-cut experimental designs, such as X irradiation at lethal and supralethal levels, the degree of animal health required for the successful completion of the experimental work is critical. In certain other experiments, particularly the acute type, it may be argued that much less emphasis is placed on animal health. However, for purposes of this chapter, we will assume that a high degree of animal health is necessary for the experiments involved, and we will approach the discussions with that in mind. Of course, specific precautions, such as animal-testing programs, become essential when SPF, DF, and GF animals are being used. For example, the SPF animal requires constant routine monitoring to assure the absence of pathogens. However, the day-to-day health problems of a conventional mouse colony are probably the most difficult.

Table 3.1 includes the location of the mouse within the animal kingdom, hematological data, and additional physiological parameters prepared from various sources. Although all the information in the table is directly applicable to the laboratory mouse, *Mus musculus*, many of the values are strain dependent.

TESTING AND MONITORING PROGRAMS

As we discussed previously, several ecological classifications of mice must be monitored constantly to make sure that the animals continue to fulfill the requirements of the classification. For example, germfree animals must be constantly monitored for the presence of organisms, and specific-pathogen-free animals must be constantly monitored for the presence of the specified pathogens from which they are declared to be free. In addition, mice

TABLE 3.1

Basic Data on the Common Laboratory Mouse, *Mus musculus**

Phylum	Chordata	(1) Family	Muridae
Subphylum	Vertebrata	(a) Genus	*Mus*
Class	Mammalia	Species	*musculus*
Order	Rodentia	(b) Genus	*Microtus*
Suborder	Myomorpha	Species	*pennsylvanicus* (meadow and field mouse, field vole)
		(2) Family	Zapodidae (jumping mouse family with cheek pouches, and long hindlimbs and tail)
		Genus	*Zapus*
		Species	*hudsonius* (jumping mouse, kangaroo mouse)

Leukocyte Values (1000/mm^3)	GF	DF	SPF	Conv.
Total WBC count	3	3.4	3.5	6–8
Polymorphonuclear	8%	14%	14%	15–25%
Lymphocytes	92%	86%	86%	73–85%
RBC ($\times 10^6$)	5.7–7.7	6.5–7.0	7.1–7.7	7.7–12.5
Hematocrit (ml/100 ml)				41–48
Platelets (1000/cm^3)				246–339
Hemoglobin (mg/100 ml)				14.8

Daily H$_2$O consumption	4.2–6.9 cm^3
Daily food consumption	3–5 g
Average litter size	3–9
Birth weight average	0.5–1 g
Eyes open	Average 11th day
Begin to eat solid food	Average 11th day

TABLE 3.1 continued

Average adult ♀ weight	25–40 g
Average adult ♂ weight	20–40 g
Breeding life ♀	6–10 litters
Breeding life ♂	1–1.5 years
Daily urinary volume	1–2 cm^3
Gestation period	19–21 days
Age at weaning	21 days
Age at puberty	35 days
Age at mating	6–10 weeks
Breeding season	Throughout year
Sexual cycle	Polyestrus
Duration of cycle	4 days
Average estrus period	12–14 hr
Copulation occurs	Onset of estrus
Fertilization site	Fallopian tube
Fertilization time	About 2 hr after mating
Segmentation of ovum to formation of blastocele	2.5–4 days
Implantation	4–5 days
Postpartum estrus (after removing litter)	2–4 days
Chromosome number	40
Temperature (rectal)	97.5F; 37.4C
Life-span	Average 1–2 years; maximum 3+ years
Blood pressure	Systole 147; diastole 106
Heart rate (beats/min)	350–750

* Data compiled from personal observation and (1) *General Information and Biological Values for Laboratory Animals*, Animal Care Facility, Stanford Medical School, Palo Alto, California; and (2) *Physiological Data for Common Laboratory Animals*, Teklab Incorporated, Distributor of Rockland Diets, Monmouth, Illinois.

of any class that are purchased from commercial sources and handled in a receiving and quarantine facility must be tested and monitored.

By necessity, disease prevention and control hinge on adequate diagnosis. Although it is highly desirable to have your own testing and monitoring program, the service can be contracted for with commercial laboratories or, perhaps, other departments of your university or research center. For example, murine virus testing services can be purchased today by simply sending serum samples collected in your colony to a commercial laboratory (e.g., Microbiological Associates, Inc., Bethesda, Md.). This method is more economical than setting up a virus testing program of your own.

The number of animals tested from each group, and the specific pathogens tested for, require that a decision be made as to statistical significance of test numbers. To arrive at the arbitrary but useful numbers, we need much background information, particularly regarding the source of the animals.

If they come from a commercial source where there is very limited laboratory control, much more testing is needed to establish the background disease information, and an increase in testing is required after the background is established. For monitoring closed colonies under laboratory control, once the background information is established and precautions are taken against accidental contamination the level of testing can be reduced to a spot-sampling procedure, so long as each room is sampled randomly on a regular weekly basis. We feel the sample should be 2 to 5% of each room each week. Certainly economics must be taken into consideration, and cost of both the experimental design and the effect of disease incidence on that design must be considered carefully. It is insufficient to test animals that become sick during the course of an experiment, because at that point it is usually too late to do anything about the problem.

One of the first questions usually asked is which pathogens should be tested for on a routine basis and which should be left for a diagnosis of disease in sick animals. The answer to this complex question is difficult and again depends on the type of experimental design, the ecological classification of the animals, and the source of the animals. For example, animals that will undergo lethal or supralethal irradiation, or other similar stresses, must be free of *Pseudomonas* sp. if the early death syndrome is to be avoided. Many other diseases can complicate experimental procedures to greater or lesser degrees. We recommend a comprehensive monitoring system that screens the animals for a broad range of diseases—bacterial, parasitic, rickettsial, and (particularly) latent viral diseases. At the present time, routine screening procedures for mouse leukemias are not readily available, but, in time, even these will be amenable to a monitoring program.

DISEASE PREVENTION

In a laboratory mouse colony, the most important part of maintaining animal health is an efficient disease-prevention program, which, in essence, becomes a public health program for mice. Because of the difficulty of disease control in mice once a sickness has become established, and because of the economic considerations involved in individual treatments of large numbers of mice, it is essential that most disease problems be solved in advance if possible.

As discussed in Chapter 1, one method of disease prevention is the use of germfree or pathogenfree animals. However, we are here interested in disease prevention primarily in conventional facilities and in conventional mice where disease is much more likely to occur. Disease prevention can be conveniently divided into the following categories: (1) immunization; (2)

prophylactic medication; (3) sanitation; (4) isolation; and (5) test and slaughter.

In general, immunization programs are not used very successfully with mice. One major exception may be the use of human smallpox vaccine virus to protect mouse colonies against ectromelia (mouse pox). This system is often used in mouse facilities that have become infected with ectromelia, because the remainder of the colony must be protected from the epizootic and from the high mortality that may result. However, we do not recommend it as a prophylactic procedure.

Other types of immunization have been used with limited success. Several investigators have tried to immunize against *Salmonella* sp. in mice (CARNOCHAN and CUMMING, 1952; BARON and FORMAL, 1960). Others have attempted immunization with inactivated lymphocytic choriomeningitis virus (MILZER and LEVINSON, 1949). However, no immunization program for mice has become practical enough to become widely used.

Prophylactic medication, such as chlorinated water, antibiotic treatment following shipping stress, routine dipping of animals for removal of external parasites, and many other techniques, can be advantageous if the diagnostic capability is available to determine what is needed.

Sanitation involves adequate cleaning and frequent changing of cages, water bottles, and bedding (we recommend changing water bottles three times weekly and cages and bedding at least twice weekly). An important aspect of sanitation and disease prevention is the number of animals housed per cage.

Isolation techniques, such as filter-top cages, are extremely important in disease prevention. In an 18-month experiment at our laboratory, non-infected mice that were held in filter-top cages in a highly contaminated area were monitored weekly and found to remain free of disease (SIMMONS *et al.*, 1967).

The test-and-slaughter method, which is commonly used in the livestock industry for disease control, is very useful in a program to prevent mouse disease. In many ways a large mouse colony is similar to a large animal herd and presents the same logistical problems. Mice found to be contaminated with undesirable organisms are not allowed into the experimental animal facilities from the quarantine area, although they may be diverted to other uses that do not require long-term holding.

Nutritional Requirements

Many so-called "ideal" formulas for mouse diets have been developed, with all commercial companies having their own "special" formulas. Similarly, many investigators, as a result of a specific mouse dietary problem, have formulated their own unique recipes. These bizarre formulas are designed primarily for conventional mice, inasmuch as the diets are unsterilizable for use with other classes of mice. The destruction of heat-labile vitamins and resulting antimetabolites is of major concern when sterilizing feed. There are, however, vitamin-fortified diets available that can be sterilized. The need for sterilizing feeds has added to the problem of formulating a completely adequate diet. BLEBY (1967) and WARD (1967) have recently discussed the effects of sterilization on nutritive value. In general, mouse diets have been formulated according to general usage in the colony—that is, growth, reproduction, and lactation.

In this section, we consider the nutritional requirements of mice. Commercial diets vary as a result of the ever-changing source and quality of the ingredients. The analysis attached to each bag of commercial feed therefore, can be referred to only when we make general comparisons between two different diets. The actual evaluation of a particular diet must ultimately be determined by the mouse's physiological behavior when fed this diet. A diet adequate for general growth may not be sufficient for production and lactation. Also, the important role of the diet in resistance or susceptibility to radiation, infection, drug response, and so on, has been demonstrated (SCHNEIDER, 1956, 1960; SIMON, 1960).

An important consideration concerning mouse diets is the form in which they are prepared. Usually mouse diets are pelleted, although mash (ground) and liquid forms are available. The latter form, although very difficult to work with, may be useful for germfree mice. Mash diets, when heat-sterilized, have been found to retain a greater percentage of heat-labile thiamine than pelleted diets. Also, liquid diets can be filtered with minimal loss of nutrient value.

Work at the Jackson Laboratory has shown a relationship between the maintenance of various inbred strains and specific diets (GREEN, 1966, Chap. 5). Work with defined-flora mice has shown altered production performance when compared with the same diet in germfree or conventional mice of the same strain (BRICK, 1966). Nutritional imbalance is equally important as a nutritional deficiency. Table 3.2, listing nutritional requirements, does not account for: (1) additional requirements owing to pregnancy, lactation, and so on; (2) strain variation; (3) overfortification for heat-labile vitamins; (4) changes in source of fats, proteins, and so forth, in feed composition from season to season.

DISEASE CONTROL

Theoretically it is possible to treat individual mice for any disease that has a known treatment; however, in actual practice, most treatments must be given on a large scale, with the smallest treatment unit practicable usually being the single cage unit. In most instances, this type of treatment necessitates that the drug be given with either the feed or the water as the transport vehicle. Animals may be treated individually (e.g., external parasite control), but such a procedure is extremely time-consuming and costly. In mouse colonies, the old saying "An ounce of prevention is worth a pound of cure" is certainly applicable.

In large mouse colonies, disease control is directly related to traffic control—that is, movement of personnel, such as investigators, visitors,

TABLE 3.2

Generally Accepted Nutritional Requirements for the Laboratory Mouse

Gross	
Crude protein	20–25%*
Crude fat	10–12%
Carbohydrate	45–55%
Crude fiber	4% or less
Ash	5–6%
Amino acids	
D- and L-methionine	0.39%
D- and L-phenylalanine	0.88%
L-valine	1.00%
L-leucine	1.50%
L-isoleucine	0.98%
L-threonine	0.69%
Lysine	1.05%
Cystine	0.29%
Tryptophane	0.25%
Arginine	0.85%
Histidine	0.40%
Tyrosine	0.66%
Minerals	
Co (cobalt)	0.0001–0.0003% (0.1–0.3 ppm)
I (iodine)	0.00278% (2.78 ppm)
Cu (copper)	0.014–0.015% (14–15 ppm)
Ca (calcium)	0.6–0.8% (600–800 ppm)
P (phosphorus)	0.5–0.7% (500–700 ppm)
Mn (manganese)	0.02% (20 ppm)
Na (sodium)	0.75% } or salt (0.5–1.0% of diet)
Cl (chloride)	0.95%
Fe (iron)	0.0025% (2–3 ppm)
Zn (zinc)	0.004% (4 ppm)
K (potassium)	0.2% (200 ppm)
Mg (magnesium)	0.02% (20 ppm)

TABLE 3.2 continued

Vitamins

A	250–300 IU/kg of ration
D	150 IU/kg of ration
E	20–25 mg/kg of ration
K	1 mg/kg of ration
B_1 (thiamine)	15–25 mg/kg of ration
B_2 (riboflavin)	5 mg/kg of ration
Niacin	25 mg/kg of ration
B_6 (pyridoxine)	1–5 mg/kg of ration
B_{12}	4–5 µg/kg of ration
Pantothenic acid	6–7 mg/kg of ration
Biotin	20–50 µg/kg of ration
Folic acid	0.5–1.5 mg/kg of ration
Choline	1 g/kg of ration
Inositol	50–100 mg/kg of ration

* Units are given as portion of total ration.

Sources of Information on Nutritional Requirements in Table 3.2

Bell, J. M. "Nutritional Requirements of the Laboratory Mouse." *Report of the Committee on Animal Nutrition*, National Academy of Sciences—National Research Council, Publication 990, 1962.

Morris, H. P. "Review of the Nutritive Requirements of Normal Mice for Growth, Maintenance, Reproduction, and Lactation." *J. Natl. Cancer Inst.* **5**: 115–142, 1944.

Personal observation and use of acceptable commercially prepared diets.

maintenance personnel, and, above all, animal caretakers throughout the animal quarters. As a standing policy in our animal areas, only personnel involved with the animals are allowed routine access to the area. All other personnel require permission in advance. Also, personnel working with conventional mice are not allowed to work in barrier areas on the same day.

The animal caretakers are provided with coveralls that are to be changed daily. Caretakers with a history of staph infections are not permitted to work in barrier areas. On occasion, caretakers with mild upper respiratory infections have been required to wear a face mask while working with animals in the conventional facilities. Routine nasal and rectal swabs of the caretakers may help to control undesirable contamination and spread of certain organisms. As a matter of routine policy, we require and provide time at the end of each day for all the animal caretakers to shower before going home.

A very important consideration, often overlooked in many animal colonies, is the problem of pest control. The word "pest" as used here encompasses such undesirable creatures as cockroaches, silverfish, houseflies, wild rodents, birds, and bats. Although many present-day animal facilities are built to prevent the entry of any of the above, it is still a good idea to maintain a regular control program.

A most important consideration is the elimination of human food (caretaker lunches) from the animal facilities. Lunch rooms should be located as far away from animal floors and mouse rooms as possible. Fine screen or wire mesh over all outside openings will prevent the entry of flies, birds, etc. Bait traps or electrical fly traps, located near frequently opened outside doors, will help to reduce the number of flies during warm weather. The use of door seals and silicone or putty to seal around conduits, pipes, and wires entering animal rooms will aid in preventing pests from entering these areas. Frequent spraying of stairwells, elevators, outside storage bins, hallways around mouse rooms, and so on, with Korlan 24E (Appendix) will help to maintain control of crawling insects, such as cockroaches and silverfish.

Many undesirable parasites, bacteria, and viruses may be carried by wild rodents and birds; therefore, great care must be taken to prevent their entry. One must also consider the location and protection of stored feed and bedding used in animal colonies. Wild rodents and birds having uncontrolled access to these areas will eventually cause widespread contamination. Also to be considered as undesirable in storage areas are beetles, mites, and moths.

In an experiment conducted by SYVERTON and FISCHER (1950), it was found that the American cockroach, *Periplaneta americana*, when fed a single meal containing the virus of mouse encephalomyelitis, excreted sufficient virus to kill test mice for up to a 7-day period. Cockroaches, of course, do not cause all mouse encephalomyelitis epidemics, but pests of this nature obviously must be considered and controlled. A report by BUTENKO (1958) discusses the possibility of bedbugs being vectors of an active leukaemic agent that produces leukosis in mice. A report by MILZER (1942) indicates that the lymphocytic choriomeningitis virus is capable of survival within certain arthropods for varying periods of time. By various experimental routes, the arthropods were capable of transmitting the infection to test animals.

DISEASE IN LABORATORY MICE

Psychological factors in the susceptibility to disease in humans and animals are extremely important. They may be doubly important in artificially controlled ecological situations. We often hear the term "environmental control" used in relation to housing for laboratory animals. We need to know much more about the effects of this environmental control, on both animals with disease and those free of it.

ADLER (1965) has shown that group-housed rats are more susceptible to immobilization-induced gastric ulcers than individually housed animals. Soon after reaching captivity, wild animals exhibit gastric ulceration and death that has no known relationship to infectious disease. (We have seen this phenomenon in our own work with wild rabbits.) To study specific psychic causes for these and many other divergent findings, one would have to invest much time and effort, and more money than is generally available. However, one may speculate that the cause is probably a psychological stress acting through the pituitary and adrenal glands and causing a psychogenic hormone imbalance. At the present time, work is underway to determine the effects of housing on glomerulonephritis in the mouse (Dr. Bruce Welch, University of Tennessee Memorial Research Center, personal communication).

The important aspect is that we constantly remain aware that our experimental results can be influenced in many ways and that psychological factors may contribute at certain times to variability. ADLER (1967) has reported an excellent study on stress in animals.

In a discussion of disesases in laboratory animals, the question usually arises as to the effects of specific diseases in the laboratory animals on the progress and findings of the experiment itself. Some effects are obvious. For example, investigators studying biochemical and carcinogenic changes in aging animals are unable to do so effectively if 90% of their experimental group dies of chronic respiratory disease before reaching middle age. In studies involving ionizing radiation, the now classic "early death syndrome" is known to be caused by the radiation's effect on latent *Pseudomonas aeruginosa*. However, many of the effects of laboratory animal disease on experiments are much more subtle. In this laboratory, spontaneous leukemia in aged $BC3F_1$ hybrids and spontaneous lymphocytic leukemia in caesarian-derived rats (both types from commercial sources) presented difficulties in interpretation of experimental data (unpublished observations, Richter and Albright, Richter and Simmons).

There are many reports of diseases of mice interfering with an experiment in progress. There is always a danger of a latent polyoma virus being activated on serial transplants, or being introduced into a susceptible mouse with a stock tumor (some of which are infected with latent murine viruses).

Individual diseases are discussed in the following categories: noninfectious, infectious, parasitic, experimental, genetic, neoplastic, and nutritional. Where some diseases obviously fall into multiple categories—for example, murine viral leukemias—we have arbitrarily placed them in the category we think most appropriate. The importance of some of the diseases as biomedical models for studying human disease counterparts should not be overlooked.

Noninfectious Disease

SPONTANEOUS POLYARTERITIS

Cardiovascular disease is assuming increasing importance in medicine as a primary cause of mortality in aging humans. Cardiovascular diseases in the small experimental animals have not been extensively studied. The occurrence of periarteritis in aging rats, and necrotizing polyarteritis and coronary arteritis in mice, may provide experimental models for study of various spontaneous cardiovascular lesions of man of particular interest to pathologists and biochemists.

LOWENTHAL (1927) described necrotizing aortitis and aortic aneurysm in an aging mouse, and commented on the similarity of these lesions to periarteritis nodosa of man. KUHL (1956) described periarteritis nodosalike lesions in the "Reims" outbred stock of mice following β-naphthylamin injection. FAPPENHEIMER (1958) reported pulmonary arteritis in mice infected by a rickettsialike agent. STEWART et al. (1958) observed gastric arteritis following injection of 20-methylcholanthrene into the stomach wall of C57BL mice. DERINGER (1959) reported spontaneous necrotizing polyarteritis associated with ovarian and renal amyloidosis in aged BL/De females. BALL and coworkers (1963, 1965) studied spontaneous and dietary-induced cardiovascular lesions in DBA and Swiss mice. UPTON et al. (1967) reported the occurrence of necrotizing polyarteritis in aging mice of the RF stock.

Numerous etiological factors are implicated in cardiovascular diseases of aging animals. Human and veterinary medical researchers have suggested dietary lipids and/or cholesterol, genetic predisposition, endocrine imbalance, infectious agents, and "wear-and-tear" injuries of aging as primary causes. Perhaps some combination of these factors is the cause.

Necrotizing polyarteritis in mice is related to aging. The disease apparently begins as a periarteritis, and the inflammation extends into the medial and intimal tunics. Arteries most frequently involved include the mesenterics, renal, aorta, coronaries, and utero-ovarians. In general, the descriptions of the lesions of arteritis by various investigators agree. In the mouse, arteritis is not detectable grossly. Development of the microscopic lesions is described by UPTON et al. (1967). The first indication of the lesion is a localized perivascular infiltration of nonsuppurative leukocytic elements, primarily lymphocytes. As the lesions develop, the media and intima are invaded by the inflammatory cells. Presence of the inflammatory exudate in the media often is associated with necrotic changes leading to weakening of the arterial wall and eventually an aneurysm. Rupture may occur, producing hemorrhage and, if the artery involved is a large one, death by exsanguination. If the mouse survives the acute inflammatory stages, the nonsuppurative exudate is replaced by fibroblastic infiltration and fibrosis, with obliteration of the arterial wall by the scar. Calcification is sometimes seen. In arterioles, hyaline degeneration and necrosis rather than fibrosis may occur. The actual incidence of arteritis in aging mice remains unknown. It seems probable that it is more common than has been realized and that it is also apparently strain related. The relationship of immunological factors, renal lesions, and other possible etiological factors to the pathogenesis of the periarteritis–polyarteritis syndrome of the aged mouse is not well understood.

CHLOROFORM TOXICITY IN MALE INBRED MICE

Mature male mice of some inbred strains, including DBA and C3H, are highly susceptible to low levels of chloroform in the air; thus chloroform should be considered as a possible cause of death if large numbers of mature male mice die while nearby females and immature males are unaffected. Needless to say, chloroform should not be used within mouse colonies unless absolutely necessary, and then only if properly ventilated with an exhaust hood. The dropping of a jar of chloroform in a mouse room, even though cleaned up immediately, has caused high mortality in a local colony of mice. The pathology appears to be centered primarily in the kidney (DERINGER et al., 1953). Although chloroform has a long-standing reputation as a liver toxin, the kidneys may show a white spotting on gross autopsy. Histologically, the convoluted tubules show extensive necrosis. There is no practical treatment. If a chloroform spill does occur, the nearby animals should be evacuated immediately.

HEMORRHAGIC DIATHESIS

Several investigators have reported another disease specific to adult male mice (MEIER et al., 1962; ANGEVINE and FURTH, 1943), which is characterized by severe hemorrhage into pleural and pericardial cavities and into the testes. It is usually accompanied by a severe myocarditis. Although considerable work has been done on this condition and the causative factors of the sudden death seem quite clear, these factors are not really well understood. One cause is apparently ethylene glycol precipitation in wood-shaving bedding that has been sterilized previously by the ethylene oxide–carbon dioxide method; however, this disease also occurs without this well-defined causation. For example, MEIER et al. (1962) report a "prothrombin complex" deficiency in animals suffering from this disease.

AMYLOIDOSIS

The deposition of amyloid (white, insoluble, starchlike protein) in virtually any organ in the mouse is called amyloidosis. It can be caused by many different stress factors and is irreversible (HALL et al., 1960). If it occurs in experimental animals in a high percentage, the effects of the experimental procedures should be evaluated in relation to strain of mouse used, nutrition, and methodology. This disease or condition may interefere with interpretation of experimental changes in mice and may often occur as a common finding in aging mice.

INFARCTION

Anemic or hemorrhagic infarcts may occur spontaneously in all mouse organs. The most probable cause would be occlusion of an artery caused by thrombosis. In some cases, the organ may undergo torsion and cause the infarct. Splenic arteries can be ligated to demonstrate infarcts. This condition, under ordinary circumstances, would not be considered a problem for a mouse colony.

HYPERVOLEMIA

Hypervolemia is thought to be an increase in total blood plasma. It increases blood pressure and causes pooling of blood within organs, such as the sinusoids of the liver. The viscera become congested with blood. The cause of the disease is unknown but may be associated with certain tumors (FURTH and MOSHMAN, 1951).

KIDNEY DISEASES

Hydronephrosis and polycystic kidneys occur frequently in inbred mice. The apparent increasing incidence in certain strains (e.g., C57BL/6) makes it seem probable that a genetic predisposition may be present. Many times the condition is caused by mechanical blockage of ureter or urethra causing back-pressures that result in the lesion observed. However, GRUNEBERG (1952) described a hydronephrosis as a congenital anomaly, and others (e.g., RUPPLE, 1955) have described congenital polycystic kidneys in the DA strains of mice. The disease may interfere with experimental results, particularly when the experimental groups contain small numbers of animals or involve toxicity studies. Careful autopsy procedures should make diagnosis routine.

RINGTAIL-LIKE DISEASE

This extremely troublesome condition of mice may be associated with improper environmental control. The disease often occurs in mice that have recently been shipped from one part of the country to another. The tail shows an annular constriction, often in more than one place. Edema occurs and eventually the tail undergoes necrosis and sloughs off. This occurrence follows a process similar to some amphibian metamorphoses except that here it is pathological. FLYNN (1959) discusses this disease in some detail in rats. We do not feel that the disease is amenable to treatment; however, humidity (50%) control and good nutrition, both prior to and after shipment, may be beneficial.

HEPATIC FIBROSIS IN AGING MICE

A recently reported disease in aging mice is hepatic fibrosis (HINTON et al., 1968), a spontaneously occurring disease in four inbred strains (C57BL, DBA, RF, and TS) fed on normal mouse diets.

MISCELLANEOUS DISEASES

In the open literature, there are numerous other noninfectious diseases of mice. Many are too obscure or too rare to mention here. A few of possible interest are colonic intussusception (HOLLANDER, 1959), tetraethyl lead poisoning (NODA, 1958), *Aspergillus flavus* toxin (WILSON and WILSON, 1964), toxic tremors (FRY et al., 1960), and an unusual spontaneous cardiac lesion of unknown etiology (FRY et al., 1960).

Infectious Disease

CATARRH

Although this disease is much more prevalent in rats, cross contamination may occur in animal facilities that house both rats and mice. NELSON (1937) made a thorough study of a respiratory disease syndrome. In the original study he referred to the disease as "infectious catarrh," and implicated *Streptobacillus moniliformis* and PPLO (pleuro-pneumonia-like organisms) in its etiology. Later (1946), NELSON referred to the disease as "chronic respiratory disease of mice and rat" (CRD), and the role of *S. moniliformis* was established to be a secondary one.

The literature is quite confusing on the subjects of infectious catarrh, chronic respiratory disease, snuffles, enzootic bronchiectasis, and chronic murine pneumonia.

The situation may be further complicated by concurrent infection with pneumonia virus and Sendai virus, both of which may cause a pneumonitis in mice under certain conditions.

Infectious catarrh symptomatology is usually a peculiar respiratory noise, termed "chattering" by NELSON (1937a,b,c). The sound resembles the gentle clicking together of one's teeth, and apparently originates from the deeper portion of the bronchial tree. Coincident with or immediately following the onset of chattering there may be a nondischarging mucopurulent rhinitis with snuffling and frequent rubbing of the nose with the front paws.

Other symptoms of infectious catarrh vary with the localization of the organism and the course of the individual case. Mice with lung infections usually appear healthy for 2 months or more. Some survive the infection for more than a year. The clinical disease appears as a loss of condition, roughened coat, and shallow, rapid respiration. Death usually follows a 3- to 5-week course of increasing anorexia, cachexia, and respiratory distress. A few mice display the clinical signs within a few days of infection, dying 3 to 5 weeks later. The chattering sound becomes more pronounced and regular as the disease progresses.

If the *Mycoplasma* has infected the middle ear, a purulent otitis media ensues. This condition is distressing to the animal but is seldom fatal. Clinical signs are the same as any otitis media, consisting mainly of the mouse's response to the pain. Otitis interna follows the establishment of the pathogen in the middle ear resulting in torticollis, head-tilt, or circling. This condition may be rapidly fatal, for the loss of balance progresses until the mouse is incapable of normal activities.

The gross pathology of infectious catarrh is helpful in diagnosis but is not definitive. The resulting purulent otitis media or interna, mucopurulent rhinitis, or chronic pneumonitis may be produced by a large variety of pathogens. Isolation of PPLO's in cultures from affected tissues provides the only definitive diagnosis.

The presence of pus in the nasal passages and middle ear is revealed by aspiration with a capillary pipette. Rhinitis is present in many of the infected animals. Otitis media is nearly as universal, but inner ear infection is relatively rare. Pneumonia is seen in some of the infected mice; it progresses slowly, usually not reaching the point of complete consolidation of a lobe. Often the only gross lesions observed are small foci of gray hepatization at the bronchial hilus of each lobe.

Microscopically, the disease produces neutrophilic exudate in the affected areas. In the lung, the bronchi are occluded by purulent exudate, and alveolar spaces of the affected lobes may be obliterated by exudate consisting of leukocytes, epithelioid cells, and debris. Frank abscess formation is rare. Lymphoid elements are pronounced and hyperplastic, particularly in the alveolar septa. Considerable fibrosis may be seen in older consolidated areas. Lymphoid elements may protrude into the bronchi, producing bronchostenosis. Vesicular emphysema may develop.

Infectious catarrh has been a problem for numerous rodent breeding colonies. The disease is widely distributed in nature and in laboratories. NELSON and COLLINS (1961) demonstrated that CRD-free colonies of mice are feasible. The disease was eliminated by hysterectomy derivation and rigid isolation of a colony. Because the disease, so far as is known, is spread only by nasal aerosols, filter tops should be very effective in halting horizontal transmission.

No acquired immunity to infectious catarrh has been demonstrated. Therapeutic measures are ineffective.

PRESUMPTIVE PPLO DIAGNOSIS

(1) Using aseptic techniques, remove the lungs and grind with tissue grinder or mortar and pestle.

(2) Inoculate a plate of Nelson agar with pestle.

(3) Skin out the head and decapitate the animal.

(4) Cut away the lower jaw and any excess muscle covering the tympanic bullae. Using sterile sharp-tipped scissors or pointed probe, puncture the tympanic bones exposing the middle ear.

(5) Withdraw any fluid within the middle ear with capillary pipette.

(6) Inoculate this fluid into a tube containing 0.5 ml of 0.01 M K_2HPO_4 and mix well.

(7) Pour mixture onto a plate of Nelson agar and incubate at 37C under 10% CO_2 tension for 7 days.

(8) Remove from incubator and examine for small typical colonies under microscope, using 430X.

(9) If PPLO-like colonies are observed, place a coverslip covered with Dienes stain upon the surface of the agar plate. Repeat microscopic examination. The PPLO colonies take up the stain, turning blue.

Although it has been shown that mice can be experimentally infected with the chronic murine pneumonia virus, the spontaneous disease is extremely rare, if it occurs at all. In an examination of hundreds of aged mice from colonies housed in the same facility with infected rats, not a single mouse showed either clinical or postmortem signs that indicated any infection with chronic murine pneumonia virus (CMPV). Mice appear to be much more resistant than rats to this infection.

MISCELLANEOUS MYCOPLASMAL INFECTIONS (PPLO)

In addition to the respiratory infections caused by PPLO, infections of the female genital tract, conjunctivae, and joints may also occur. In breeding inefficiency, where nutritional problems have been eliminated, it would probably be a good idea to check for PPLO, although no treatment has proved very successful.

SEPTIC ARTHRITIS

A number of different bacteria may cause inflammation and sepsis of the joints in mice. A specific report of *Streptobacillus moniliformis* has been published by FREUNDT (1959). Apparently a systemic infection with the organism is followed by a localization of the organism in the joints of surviving mice. *Mycoplasma arthriditis* also localizes in joints of mice, and *Corynebacterium kutscheri* has been implicated as well (ITO and TANAKA, 1952). The sequelae to any joint infection may be necrosis and eventually ankylosis. If the condition is diagnosed clinically by swelling and painful movement of the mice, broad spectrum antibiotic therapy may be beneficial, depending, of course, on the susceptibility of the infectious agent.

TYZZER'S DISEASE

This disease, first described by TYZZER (1917), is difficult to control in laboratory mice. Acute cases cause death, and chronic cases show prolonged loss of condition.

Tyzzer's disease is attributed to a gram-negative bacillus designated *Bacillus piliformis*. The taxonomic position of this organism and its exact role in the etiology of Tyzzer's disease are obscure, because the organisms cannot readily be cultivated on laboratory media and tissue suspensions containing it will not readily produce the disease.

The disease is one of the major causes of loss among laboratory mice in Europe and Japan. Since the original description, outbreaks have been reported from laboratories all over the world, yet few adequate accounts of the disease have been published. Probably an oral route of infection is normal but mice have also been artificially infected by the intravenous injection of infected material. Infection is usually enzootic, with occasional epizootic outbreaks. Once the infection is introduced, faulty management, such as overcrowding, favors epizootics; whereas, with good management, few cases of overt disease appear. Wild mice may be responsible for fresh introductions of the disease.

The disease is more often seen as an infection of mice in long-term experiments, or in mice subjected to irradiation. Frequently there are no signs of the disease in the cage mates of a mouse found dead or suffering from this infection. In an acute form that occurs frequently among newly weaned mice, severe diarrhea is prevalent for a few days prior to death. At autopsy, reddening of the intestinal mucosa, particularly at the ileocecal junction, and liver lesions are found, although these do not often occur in the most severe cases. In very acute cases, the animal develops diarrhea, passing water and slimy fecal material, and dies within 24 to 48 hr. In more chronic cases and in those with only a mild diarrhea, the liver is the only organ to show lesions. It may show very few or be almost entirely covered by white or yellowish circular areas of necrosis, some of which may show a darker central region. Histologically, infected livers show patches of necrotic tissue with extensive infiltration of polymorphonuclear leukocytes. Long, slender bacilli lying parallel to one another in bundles are frequently seen inside whole cells lying at the edge of the necrotic regions. These bacilli characteristically stain in an irregular fashion, and swollen structures, which Tyzzer took to be normal eubacterial spores, can sometimes be seen. Occasionally, adult mice of various ages are found dead, and at autopsy they show few symptoms except perhaps the presence of slightly liquid fecal material.

The typical appearance of the hepatic foci is usually sufficient for diagnosis. The slender gram-negative bacilli can readily be seen in stained sections of the liver and the affected areas of the gut. The macroscopic lesions can be confused with those caused by *Salmonella* or *Corynebacterium* but can be differentiated by successful cultivation of causative bacteria.

Tyzzer's disease is a hepatitis with intestinal involvement. In epidemic outbreaks, the acute form is that most often seen. Animals appearing to be in perfect health one day are dead the next. Slightly more chronic cases may show loss of weight and a general deterioration in condition, with the usual staring coat and a humped-back, listless appearance. There is often a diarrhea of varying severity, with death following fairly rapidly in young mice and between 2 and 3 weeks later in mice 5 weeks old or more. Chronic cases show prolonged loss of condition but very often remain undetected or unsuspected until autopsy.

Outbreaks of Tyzzer's disease are difficult to control because the disease as a whole is so little understood. In practice, however, it is possible to keep a stock of animals free from infection by means of normal hygienic measures and by rigid restriction of the introduction of stock from other colonies. Mice with diarrhea should be killed and the liver and gut carefully examined. Mice in contact with sick animals should be sacrificed, the equipment should be sterilized, and the bedding destroyed.

The administration of Terramycin in the drinking water for a 4 to 5-day period may be effective, which is the usual procedure whenever an outbreak is suspected (50 mg per $1\frac{1}{2}$ pints of drinking water).

Recommended Reading on Tyzzer's Disease

Cook, R. "Common Diseases of Laboratory Animals," in *The A.T.A. Manual of Laboratory Animal Practice and Techniques*, ed. by D. J. Short and D. P. Woodnott (Springfield, Illinois: Charles C Thomas, Publisher, 1963), pp. 118–119, 122.

Innes, J. R. M., and A. A. Tuffery. "Diseases of Laboratory Mice and Rats " in

Animals for Research: Principles of Breeding and Management, ed. by W. Lane-Petter (New York: Academic Press, 1963), pp. 53–54, 63–64.

Kraft, Lisbeth M. "Two Viruses Causing Diarrhoea in Infant Mice," in *The Problems of Laboratory Animal Disease*, ed. by R. J. C. Harris (New York: Academic Press, 1962), pp. 115–130.

Worden, A. N., and W. Lane-Petter, eds. *The UFAW Handbook on the Care and Management of Laboratory Animals*, 2nd ed. (London: The Universities Federation for Animal Welfare, 1957), pp. 266, 268.

CATARRHAL ENTEROCOLITIS

This disease has been reported as being associated with *Citrobacter freundii* in laboratory mice. The main clinical sign is diarrhea (BRENNAN et al., 1965). Although we have never experienced a problem with this organism, screening of a colony is recommended if such a disease syndrome occurs. Autopsy of animals with this disease shows an accumulation of mucus in the dilated colon, the wall of which is thickened and congested. Affected animals are unable to form fecal material into a solid or semisolid mass, and as a result, the hindquarters and tail are always covered with semiliquid to liquid material.

MOUSE LEPROSY

Mouse leprosy is caused by *Mycobacterium leprae murium* which has been reported to be a spontaneously occurring, acid-fast bacillus producing a leprosylike disease in the brown mouse (wild), from which the disease can be passed to the laboratory mouse (KRAKOWER and GONZALEZ, 1937). Mice have also been used for experimental maintenance in studies of the acid-fast bacillus of human leprosy (see Experimental Diseases, p. 101). It seems very unlikely that this disease could occur spontaneously in a laboratory mouse colony.

PSEUDOMONAS INFECTION OF MICE

Pseudomonas species previously considered unimportant normal flora should in our opinion always be considered potential pathogens and indicators of sanitation levels in mouse colonies. When some species of *Pseudomonas* are present in mice receiving acute lethal whole-body radiation, very early deaths frequently occur preventing use of these animals in radiation protection and recovery experiments. We define this early death syndrome occurrence as animals dying in the first 8 days after lethal or supralethal X irradiation usually brought about by undiagnosed infection of the animals with *Pseudomonas* sp. The lethal effects of *P. aeruginosa* infection in X-irradiated animals have been described by MILLER *et al.* (1952) and WENSINCK *et al.* (1957).

The presence of *Pseudomonas* in the blood stream may be determined by culturing heart blood clots from animals dying during the first 8 days after irradiation or those near death. Samples of heart blood and of minced intestines are placed in thioglycolate broth and incubated at 37C for 48 hr. At the end of the incubation period, the broth culture is streaked on glycerol or "Tech" agar plates (KING *et al.*, 1954) and incubated at 25C for 48 hr. After the incubation period, the plates are examined for the presence of pyocyanin.

It has been shown (SIMMONS and FRANKLIN, 1967) that the nonpigmented *Pseudomonas*, *P. stutzeri*, is not involved in the early death syndrome. Although nonpigmented *Pseudomonas* has not been incriminated in the early death syndrome as far as we know, it may sometimes be isolated.

Pseudomonas organisms tend to thrive in moist or damp places. They can be transmitted by fomites, dust, caretakers, or animal contact. The best method we know to keep the early death syndrome to a minimum is to use a screening program on all mice destined for use in such experiments and to sacrifice any positive animals or divert them to other uses. The methods we use for diagnosis and culturing *Pseudomonas* sp. are listed as follows:

(1) Collect approximately 10 fecal pellets from a clean mouse cage.
(2) To the sample of fecal pellets add 10 ml of tetrathionate broth.
(3) Macerate the fecal pellets in the broth with a sterile tongue depressor.
(4) Using a pipette, inoculate a tube of tetrathionate broth. Incubate the plain tetrathionate broth tube for 48 hr at 37C.
(5) After the incubation period, transfer one loopful of the culture to a plate of glycerol agar. Streak for isolation.
(6) Incubate the glycerol plate for 48 hr at 25C.
(7) Observe for pigment production.

Any pigment-producing organism is presumed to be an early-death-type *Pseudomonas* and is considered positive.

Prevention and control are based on the use of chlorinated drinking water at a level of between 10 and 16 ppm. The exact level depends upon how long it will be before the next water bottle change, and on what degree of organic matter contamination and resultant fall-off of chlorine content will occur. It is an excellent idea occasionally to do a second-phase screening of animals that have passed through the quarantine area with a clean bill of health. Apparently occasional "periodic shedders" will be missed and can be picked up in a second testing.

LEPTOSPIROSIS

Leptospirosis in laboratory mice is exceedingly rare; however, it has been reported in one colony of Swiss albino mice. The isolated agent was *Leptospira ballum* (STOENNER *et al.*, 1958). The disease was subclinical and the organism was discovered during experimental procedures. The *Leptospira ballum* was eliminated from the colony by feeding pelleted feed containing 1000 g chlortetracycline per ton for 10 days and transferring the mice on the seventh day to sterilized cages and water bottles. This instance is yet another example of why it is essential to keep pests (insects, wild rodents, etc.) from entering experimental animal areas.

MOUSE TYPHOID OR SALMONELLOSIS IN MICE

In past years, *Salmonella* species, including *S. typhimurium*, *S. enteritidis*, and others, have constituted a serious problem in laboratory mice. Such was particularly true for commercial breeders of laboratory mice. One large shipment of mice that we received in 1963 was 88% positive for *Salmonella* sp. Fecal culture was used as the method of diagnosis. The animals exhibited a large range of clinical symptoms from a perfectly normal appearance to severe diarrhea. Ordinarily they appeared normal; however, they rapidly became clinically sick after being stressed by some experimental technique, particularly X irradiation.

SHECHMEISTER (1967) and SHECHMEISTER *et al.* (1953) have shown interesting relationships of *S. enteritidis* to X irradiation. Death times with lethal doses of *S. enteritidis* are much shorter in X-irradiated mice, and such mice are more susceptible to infection with *S. enteritidis*. This fact bears out much other experience with mice that were latent carriers of *Salmonella*. Apparently adults pass the *S. enteritidis* infection to their offspring, perpetuating the disease generation after generation. Again, as with *Pseudomonas*, the only effective method of eliminating the organism is by eliminating the mice shown, by test, to be infected. Most commercial suppliers have done so, with the encouraging result that today only occasionally do we isolate a *Salmonella* in our screening program. Our diagnostic method is as follows:

(1) To the sample of fecal pellets, add 10 ml of plain tetrathionate broth.
(2) Macerate the fecal pellets in the broth with a sterile tongue depressor.
(3) With a sterile disposable pipette, with rubber bulb attached, transfer a portion of the fecal–tetrathionate suspension to a screw-capped tube containing 10 ml of tetrathionate broth with brilliant green and sulfathiazole additives.
(4) Incubate 48 hr at 37C.

(5) Shake the tetrathionate broth culture before streaking.

(6) Transfer and streak a loopful of 48-hr tetrathionate culture to a plate of brilliant green agar.

(7) Incubate at 37C for 18 to 24 hr.

(8) Examine the plates for typical-looking *Salmonella* colonies and pick a representative number of colonies from each plate.

(9) Pick only well-isolated colonies by touching the top of the colony with the needle.

(10) Inoculate a tube of TSI (triple sugar-iron agar) with each colony picked. Inoculate the TSI by stabbing the needle into the butt of the medium, withdrawing, and streaking the slant. Then inoculate a tube of TST (trypticase soy-tryptose) broth with the same colony picked for the TSI slant.

(11) Incubate the TSI and TST at 37C for 18 to 24 hr.

(12) After the incubation period examine the TSI medium for typical *Salmonella* reactions; for example,

    (a) H_2S production—medium turns black.
    (b) Alkaline slant—slant is red in color.
    (c) Acid and gas production in butt—butt is yellow and agar shows gas cleavage. Acid production sometimes cannot be noted owing to excess production of H_2S.

(13) Serological tests may now be run for H antigen test.

(14) To prepare the antigen, add to the TST broth culture an equal amount of 0.6% formalinized saline solution. Set the stopper firmly; invert the tube several times to mix, and allow to stand at room temperature for 1 hr to inactivate the O antigen and kill the organism.

(15) To the bottom of a serological tube, add 0.02 ml of Communicable Disease Center polyvalent H antisera diluted 1:5.

(16) Add 1 ml of the formalinized antigen solution.

(17) Incubate in a 48 to 50C water bath for 10 to 60 min, observing the tubes at 10-min intervals for a dense, cottony, suspended agglutination. Do not shake the tube because it will make the agglutination dissipate. A positive agglutination is a presumptive positive diagnosis for *Salmonella* sp. Cultures are then sent to the reference laboratory for typing.

PASTEURELLA PNEUMOTROPICA INFECTION OF MICE

Pasteurella pneumotropica has been described as occurring in mice exhibiting pneumonia and conjunctivitis. Apparently it may also cause, or at least be present in, urocystitis and metritis (BRENNAN et al., 1965). In our diagnostic program we have not found this problem to be serious; however, it should be considered if any of these problems arise, particularly in mice received from commercial sources.

TULAREMIA

Tularemia (*Pasteurella tularensis*) has occurred spontaneously in a closed hamster colony (PERMAN and BERGELAND, 1967), and could occur in closed mouse colonies as well if the source of the infection were wild rodents, food, bedding, or insect pests. Unusual disease syndromes should never be ruled out as unthinkable. PERMAN and BERGELAND's report is an excellent example of a highly unusual disease in a closed colony.

MOUSE POLIOMYELITIS OR THEILER'S MOUSE ENCEPHALOMYELITIS

Theiler's mouse encephalomyelitis is a viral disease characterized by flaccid paralysis resulting from necrosis of the ganglionic cells of the ventral horn of the spinal cord. The virus is carried in the intestinal tract of many laboratory mice. Its wide distribution and ability to produce cord lesions like those of the poliomyelitis viruses have made it of considerable interest.

The virus is usually found wherever mice are raised. All strains of the virus given intracerebrally will infect young mice, but only the more virulent strains, such as Theiler's GDVII and FA strains, readily infect mice intranasally or intraperitoneally. Intracerebrally, minimal infective doses are required for infection. Virus administered by both the intracerebral and intranasal routes results in typical encephalitis and paralysis. Virus given by the intraperitoneal route generally produces paralysis without encephalitis and with a lower mortality.

Atrophy of muscles is the only grossly apparent pathologic change. Microscopic study shows that the paralysis results from necrosis of ganglionic cells of the ventral horn. Degeneration of these cells has been observed before the onset of paralysis. The necrosis is acute and is followed by neuronophagia. There may also be some focal and perivascular infiltration of the central nervous system by mononuclear cells.

The pathologic changes are neuronal necrosis, neuronophagia, perivascular mononuclear cell infiltration, and glial cell proliferation, occurring mostly in the cord and the ventral horns.

Diagnosis is based on the following: (1) flaccid paralysis developing 7 to 37 days after intracerebral injection of the virus; (2) characteristic microscopic lesions appear in the central nervous system—that is, necrosis of ventral neurons; and (3) serology will indicate past presence of the virus.

Elimination of this infection must depend upon the founding of a GDVII-free stock of mice derived from the offspring of old breeding females that have eliminated the virus, or from young fostered on the known GDVII-free foster parents.

Inasmuch as most laboratory mice are used at the age when they are susceptible to infection of the central nervous system, care should be taken to observe each mouse individually while it is moving about to eliminate those that show evidence of paralysis or encephalitis. Mice should be examined for muscular atrophy or weakness. Questionable mice should not be used.

The high incidence of the virus in all mouse colonies examined suggests the difficulties involved in obtaining breeders from which to start a clean colony and the need for strict isolation to maintain such a colony. These difficulties have caused most laboratories to be on the alert for individual mice with central nervous system symptoms and to discard the infected mice, but otherwise to ignore the presence of the virus.

The prognosis states that most mouse colonies appear to be infected with Theiler's virus. With rare exceptions the virus is carried in the digestive tract and excreted periodically over prolonged periods of time; even so, only a few mice acquire an infection of the central nervous system. If the virus produces any recognizable pathologic change in the intestinal tract, the lesions have not been reported. Only young mice appear to be susceptible to invasion, and they either die or develop obvious and quite characteristic paralysis and show anterior horn lesions simulating those of human poliomyelitis.

Recommended Reading on Mouse Encephalitis

Holdenried, Robert, ed. *Viruses of Laboratory Rodents*, National Cancer Institute Monograph No. 20 (Bethesda: National Cancer Institute, 1966), pp. 28, 71–73.

Lane-Petter, W., ed. *Animals for Research* (New York: Academic Press, 1963), p. 66.

Maurer, Fred D. "Mouse Poliomyelitis." *J. Natl. Cancer Inst.* **20**: 871–874, 1958.

K VIRUS INFECTION

K virus was first isolated from C3H mice by KILHAM and MURPHY (1953) when it was discovered to be a complicating factor in experiments on Bittner's Jackson Laboratory mammary tumor-inducing milk agent. The virus has since been shown to be widespread in both laboratory (PARKER *et al.*, 1966) and wild (HOLT, 1959) mouse populations. The K virus was also isolated from a virus-passed mouse leukemia (ROWE *et al.*, 1962) where it was again revealed as a complicating factor in an experiment. No relationship between K virus and the oncogenic viruses was found.

On the basis of its physical characteristics, K virus is placed in the papovavirus group; however, K virus is not tumorigenic. It is a double-stranded DNA virus that is ether resistant and heat stable.

The K virus is pantropic, although the pathological lesions produced are primarily in the lungs. The virus produces a fatal pneumonitis associated with viremia in suckling mice, but is nonpathogenic in the adult. Experimental infection can be attained by inoculation of the virus by intranasal, subcutaneous, intraperitoneal, intracranial, or oral routes. The isolation of K virus from the mammary glands of nursing female mice suggests possible transmission from dam to suckling young through milk. Infective virus is also shed in the urine.

Resistance to pathogenic effects of K virus begins at 12 days of age and is practically completed by 18 days. Mice rarely survive K virus infection at 8 days of age (KILHAM and MURPHY, 1953).

Infection of young suckling mice produces no overt disease signs for 7 to 9 days. The infant mice grow normally and are well-nourished until the disease becomes manifest on the eighth or ninth day. The course of the clinical disease is peracute, death intervening about 5 hr after the first signs of distress. Moribund animals still appear to be in good condition, with ingested milk present in the stomach.

The principal clinical sign of the disease is a peculiar forced respiration described as "chugging" (FISHER and KILHAM, 1953). Progressive cyanosis is sometimes seen. The abrupt onset and rapidly fatal course of the disease make clinical diagnosis of doubtful value.

The gross pathology of K virus infection is diagnostically unreliable. Pulmonary lesions are grossly visible in only 50% of suckling mice dying from the disease, probably owing to the rapid development of the peracute phase of infection. In those animals displaying gross pulmonary lesions, the distribution of the lesions is patchy, and typical stages of pneumonia are sometimes seen in the larger lesions.

The microscopic pathology of the K virus pneumonia is unique among the described murine pneumotropic viruses and is constant in infected mice. The pathognomic lesion is enlargement of the endothelial cells of the pulmonary blood vessels with a characteristic enlargement and vacuolation of the nucleus, termed "ballooning" (FISHER and KILHAM, 1953). A Feulgen-positive homogenous inclusion body, 4 to 8 μ in diameter, appears in the clear areas of the nucleus. The 2 to 4 × enlargement of endothelial cells occludes some of the smaller vessels, particularly the septal vessels.

The lung tissue shows a diffuse nonpurulent interstitial pneumonitis. The alveoli are mostly clear; a few contain acellular serous exudate. No pleuritis or bronchitis is seen.

The presence of K virus in a colony may be detectable by the demonstration of either hemagglutination inhibition (HI) or complement fixation antibodies. Virus may be demonstrable by intracranial inoculation of newborn mice with suspect tissue, or use of the intranasal mouse antibody production test. The HI and newborn inoculation tests are more sensitive.

The virus has not been shown to cross the placental barrier. Therefore, hysterectomy derivation of the colony should eliminate it. Use of filter tops with a culling program may also be effective in checking the spread of K virus in a mouse population.

REOVIRUS TYPE 3 INFECTION

Reovirus type 3 is a ubiquitous agent possibly associated with pneumonia, steatorrhea, alopecia, hepatitis, and encephalitis in man, but more commonly isolated from healthy individuals. Natural epizootics as well as artificial infection of mice produce some or all of the same clinical signs. This virus group also has been isolated from other species of mammals.

In 1953, STANLEY et al. recovered an agent they called hepatoencephalomyelitis virus (HEM) from a child. This virus produced a characteristic "oily hair effect" in suckling mice. At that time, these authors reported that Dr. E. L. French had isolated a similar agent, murine encephalomyocarditis virus (MEM), from a colony of laboratory mice. Both agents were later established to be a Reovirus type 3 by hemagglutination inhibition typing.

COOK (1963) reported successive epizootics in his mouse colony owing to an endemic agent identified as MEM virus. Cook's mice were derived from French's colony in 1949. Severe epizootics occurred in September 1961 and February 1962. Reovirus type 3 of both human and mouse derivation is easily transmitted among laboratory mice by oral, and probably by respiratory, infection or by intraperitoneal or intracerebral injection.

Cook reported the naturally occurring infection in his colony to be endemic and usually asymptomatic. Scattered incidents of affected litters were seen until the first major outbreak of overt infection when 130 of 800 litters were infected. Clinical signs of diarrhea, jaundice, oily skin and hair, and severe stunting usually occurred in first litters, when the young mice were 10 to 14 days old. The parents of infected litters were observed to have oily hair. From one to five young mice were stunted and icteric in an infected litter. The remainder of the young displayed oily hair and sometimes mild diarrhea.

Encephalitis first appears about the tenth day, with degenerative changes in the neurones and perivascular lymphocytic cuffing. Ecchymotic hemorrhage and frank necrosis occur by day 14, with a widespread and severe encephalitis. Other lesions—such as pulmonary alveolar hemorrhage and edema, follicular hyperplasia of the spleen, and suppurative to nonsuppurative dermatitis—are less constant findings. Interstitial pneumonia occurs in most affected mice through the tenth week. Other organs appear essentially normal.

The lesions of the early acute phase of Reovirus type 3 infection would seem to result from direct virus-induced cell damage (WALTERS et al., 1963).

The secondary clinical signs follow logically—icterus, incoordination, and steatorrhea with the oily hair effect, evolving from the liver, brain, and pancreatic necrosis.

The abrupt alteration in the disease process that occurs at the fifth week marks the transition from the acute to the chronic stages. Although it coincides with puberty in the mice, the relationship is unclear. A part of the pathology of the chronic form of this disease apparently results from the earlier tissue damage, but STANLEY *et al.* (1953) do not feel that this explanation is consistent with their observations of the disease.

COOK (1963) showed that HEM and MEM were serologically and pathogenically identical, suggesting possible health hazards from both humans and mice in the laboratory. STANLEY *et al.* (1953) have suggested that man is the natural reservoir host for the reoviruses; if so, yet another zoonosis may be present in the laboratory colony.

Prior to and between the epizootic occurrences, overt infection was occasionally seen. Five- to six-week-old mice were sometimes observed to be icteric. Testing for HI titer showed the virus to be widespread in the colony. Antibody transfer by colostrum was found to account for the resistance of most of the first, and nearly all subsequent, litters.

Animals in the acute stages of Reovirus type 3 infection are often obese and icteric and show the characteristic oily skin and hair at necropsy. A peritoneal exudate is common; it may be stained by bile or hemorrhage. The liver is enlarged and darkened with circular hollow lesions about 3 mm in diameter scattered on all surfaces. After 7 days, the intestines appear hyperemic and distended. The heart may show small circular gray epicardial foci, and some ecchymotic pulmonary hemorrhage may be present. After 10 days the brain is swollen and congested. Pronounced icterus is present, the gall bladder is black and swollen, and the intestinal contents and urine are yellowed.

Macroscopic appearance of the chronic disease is very similar, except cachexia replaces the fat deposits seen in the acute form. Alopecia is usual, and mild splenomegaly is occasionally seen.

By the tenth day the entire pancreas has undergone necrosis, except the islets, which remain normal. Similar changes occur in the salivary glands when infection is by the oral route.

No change occurs in the heart or the skeletal muscle until the seventh day, at which time small areas of Zenker's necrosis are seen in the myocardium; these increase progressively in size and number with little inflammatory response. Fibroblastic infiltration begins about day 10. Skeletal muscle reacts similarly, with calcification of the necrotic foci beginning about day 14.

The method of transmission of Reovirus type 3 from mouse to mouse remains in some doubt. Transmission by oral routes, especially coprophagy, are probable, and transmission by bites and respiratory routes seems likely. Fortunately, the oily hair-and-skin syndrome is unique, providing a rapid tentative diagnosis.

Cook was apparently successful in controlling, but not eradicating, Reovirus type 3 by a culling program. Control of horizontal transmission by filter-top cages is feasible. Eradication is possible only by slaughter, decontamination, and beginning with clean stock. Many caesarean-derived SPF colonies seem to be free from the virus.

MURINE HEPATITIS

The mouse hepatitis viruses have not yet been classified with any of the established virus groups, but they bear at least a superficial resemblance to myxoviruses in that they replicate in the cytoplasm of the cells which they infect, are of medium size, are chloroform-sensitive viruses, and are moderately resistant to sodium deoxycholate, as would be expected from their ability to produce enteric infection. Serologically, all strains tested share common complement fixation and neutralizing antigens, but no two strains from different sources appear to be identical.

In a study of indigenous viruses of mouse colonies, it was found that mouse hepatitis was the third most commonly observed virus infecting mouse colonies. However, the number of mice within infected colonies possessing antiviral antibody was, with one exception, less than 50%. Based on this low percentage, one might conclude that the virus has a low infectiousness for mice. This conclusion is not true, for it has been shown that mouse hepatitis is a highly contagious virus that may also be isolated from fecal material during an apparently short, acute period of infection.

Weanling mice infected with mouse hepatitis virus when exposed by the intraperitoneal, intracerebral, subcutaneous, intramuscular, and intranasal routes, could not be infected per os, and only a small number became infected following intragastral inoculation. Exposure to an aerosol resulted in infection of approximately 80% of the young mice.

Similar to many other murine agents, the mouse hepatitis viruses normally occur as nonpathogenic agents that, under conditions of stress or experimental manipulation, may induce disease.

Three types of experimental stress have been particularly associated with induction of hepatitis owing to these agents: (1) infection with *Eperythrozoon coccoides*, (2) passage of mouse leukemia by transplants or viruses, and (3) neonatal thymectomy.

Recently mouse hepatitis viruses have been found to be a potentially great source of experimental error in work involved with infant mice. This problem chiefly occurs when litters from a noninfected colony are exposed to infected but asymptomatic mice, such as in experimental animal rooms housing mice from more than one source. Under these conditions, the mortality rate in the nonimmune sucklings may be as high as 50 to 75%, with death chiefly attributable to an encephalitis rather than a hepatitis. Even suckling mice of an infected strain may be involved, provided the level of environmental contamination is sufficiently high; however, the mortality rate is lower than in nonimmune animals.

The natural history of these viruses is presently thought to be as follows. In a colony of healthy, infected mice there is probably a stable epizoologic pattern. The viruses are extremely contagious, with the result that most mice acquire virus early in life, possibly before weaning.

Essentially, infection is restricted to the intestinal tract, and entry into other tissues during the acute stage of infection is probably inhibited by maternal antibody. This lack of generalized infection results in a lesser degree of antibody response than in mice infected without passive antibody protection. However, the antibody titer may be sufficient to be detected by a sensitive complement fixation procedure, and it is readily detected by plaque neutralization tests.

Mice appear to excrete virus in the feces for an indefinite period, and only infrequently can virus be isolated from the liver. The occasional presence of virus in the liver of an asymptomatic mouse may well be the result of a transitory portal viremia, with the Kupffer cells preventing access of virus to parenchymal cells. Damage to Kupffer cell function, such as by *E. coccoides* or leukemic infiltrates, may permit the involvement of parenchyma with resultant hepatitis.

The naturally occurring strains of virus are predominantly neurotropic with only a limited hepatotropism. The occurrence of hepatitis in stressed mice may also be a function of the emergence of more hepatotropic mutants (e.g., MH-1, MHV-3, and A-59 strains).

Recommended Reading on Murine Hepatitis

Calisher, Charles H., and Wallace P. Rowe. "Mouse Hepatitis, Reo-3, and the Theiler Viruses." *Natl. Cancer Inst. Monogr.* **20**: 67–69, 1966.

Hartley, Janet W., Wallace P. Rowe, Henry H. Bloom, and Horace C. Turner. "Antibodies to Mouse Hepatitis Viruses in Human Sera." *Proc. Soc. Exp. Biol. Med.* **15**: 414–418, 1964.

Parker, John C., Raymond W. Tennant, and Thomas G. Ward. "Prevalence of Viruses in Mouse Colonies." *Natl. Cancer Inst. Monogr.* **20**: 25–31, 1966.

Reubner, Boris H., J. Russell Lindsey, and Edward C. Melby, Jr. "Hepatitis and Other Spontaneous Liver Lesions of Small Experimental Animals," in *The Pathology of Laboratory Animals*, ed. by W. E. Ribelin and J. R. McCoy (Springfield, Illinois: Charles C Thomas, Publisher, 1962), pp. 160–168.

Rowe, Wallace P., J. W. Hartley, and W. I. Capps. "Mouse Hepatitis Virus Infection as a Highly Contagious, Prevalent Enteric Infection of Mice." *Proc. Soc. Exp. Biol. Med.* **112**: 161–165, 1963.

White, Roberta J., and Stewart H. Madin. "Pathogenesis of Murine Hepatitis: Route of Infection and Susceptibility of the Host." *Am. J. Vet. Res.* **25**: 1236–1240, 1964.

POLYOMA VIRUS INFECTION OF MICE

Polyoma virus of mice has been subjected to intensive study by many of the techniques of classical virology—animal inoculation, tissue culture cytopathogenicity, hemagglutination and hemagglutination inhibition, complement fixation, and neutralization reactions. The polyoma virus is included in the papovirus group.

Polyoma virus is widespread geographically, being reported in mice from the United States, Europe, and Japan; it is most prevalent in colonies housed in proximity to mice experimentally inoculated with polyoma virus. Infection is highly prevalent among wild mice, both in city tenements and in farm environments.

The virus seems to be spread primarily by excretion in the urine, with the main reservoir of virus in the case of the wild mice being contamination of permanent nesting areas. The most important fomites for spread of virus into colonies of both wild and laboratory mice may well be grain that was contaminated with mouse urine in mills and storage areas. Perhaps the most important factor influencing the epidemiology is contaminated urinary excretion of mice infected as infants.

Repeated sampling of a polyoma-infected colony showed a consistent 50% antibody incidence in mice older than 9 months, whereas 4-month-old mice showed considerable fluctuation in positivity, with the number of antibody-positive mice generally less than 15%. The virus in neonatal mice can induce a literally fantastic number and variety of tumors; however, it does not cause tumors in the mouse colonies in which it is indigenous. The reasons for this paradox lie in the very short period of life during which mice are susceptible to the oncogenic effect of infection, the ineffectiveness of inhalation infection in producing tumors, the low dosage of virus acquired in natural infections, and the effective transfer of maternal antibody to the young.

Control measures generally apply to all the indigenous murine viruses that would obviously interfere with any experimental investigation of tumor systems. Specific-pathogen-free and germfree colonies are remarkably free from indigenous mouse viruses (i.e., polyoma, K virus, mouse adenovirus, Reovirus type 3, mouse salivary gland virus, and thymic virus); except for Reovirus type 3, this factor points to the caesarean-derived, isolated mouse colony as the only logical direction. Contact with wild mice, either by direct ingression into colonies or by indirect spread through contamination of feed, bedding supplies, and personnel, must be strictly avoided. Avoidance of these contacts, combined with serologic and virologic monitoring, can go a long way toward providing the virologist and cancer research worker with animals in which he can have a high degree of confidence. Polyoma virus is used experimentally as a tumor-producing virus to study various aspects of tumor production.

Recommended Reading on Polyoma Virus of Mice

Parker, John C., Raymond W. Tennant, and Thomas G. Ward. "Prevalence of Viruses in Mouse Colonies." *Natl. Cancer Inst. Monogr.* **20**: 25–36, 1966.

Rowe, W. P., J. W. Hartley, and R. J. Huebner. "Polyoma and Other Indigenous Mouse Viruses." *Lab. Animal Care* **13**: 166–175, 1963.

Tennant, Raymond W. "Taxonomy of Murine Viruses." *Natl. Cancer Inst. Monogr.* **20**: 47–53, 1966.

PNEUMONIA VIRUS OF MICE (PVM) INFECTION

The virus was first isolated for study by HORSFALL and HAHN (1940), who found it complicating their attempts to isolate viruses from nasopharyngeal washings of humans with acute nonfluenzal respiratory diseases. About 65% of their mice were found to harbor PVM, thus invalidating the results of the primary experiment. The virus was isolated by serial lung suspension passage. Although PVM is presently unclassified, its physical characteristics most closely resemble the paramyxoviruses.

Infection with PVM is detected by the presence of hemagglutination-inhibition antibody, by isolation in primary hamster kidney cell culture, or in mice, or by intranasal mouse antibody production (MAP) test.

PARKER et al. (1966) found PVM distributed throughout the United States. Of 5284 mice tested for PVM infection, 16% were found to be positive by using HI techniques. Infected mice were found in 65% of the 34 colonies tested.

The original findings of HORSFALL and HAHN, who reported the virus as a latent infection of apparently healthy mice, have not been verified by later workers (TENNANT et al., 1966). At present, PVM is not known to produce latent or chronic-inapparent infection in mice. The dominant factor in the natural history of PVM is its low infectivity. The disease produced is an acute and frequently lethal interstitial pneumonitis. Occurrence in a colony is focal; apparently the disease is enzootic only as a reflection of the low infectivity of PVM.

The virus is strictly pneumotropic in mice, for it cannot be isolated from other tissues, and inoculations by intracerebral, intraperitoneal, intravenous, intramuscular, and subcutaneous routes fail to produce disease or recovery of the virus.

Natural transmission is apparently by intranasal aerosol inhalation; part of the low infectivity of the virus may be accounted for by the absence of copious nasal discharge by the diseased animal.

Mice infected with PVM by intranasal inoculation appear healthy for 5 to 7 days. Following this incubation period, the animals gradually display listlessness and anorexia. Both HI and complement-fixing antibody are detectable at 9 to 10 days. In the later stages of the disease, the animal is cachectic, feverish, and insensitive to its surroundings. It appears chilled, its fur is ruffled, and it tends to round up. Respiration is slow, deep, and labored, and cyanosis of the ears and tail is frequently seen. Death occurs on the twelfth or thirteenth day, caused by the combination of cachexia and hypoxia.

Macroscopic appearance of the lungs somewhat resembles influenza lesions (IIDA and BANG, 1963). Consolidation of one-half to three-quarters of the lung is seen in fatal cases. Mice killed in the terminal stages of the disease seldom show more than one-half of the lung surface affected. The lesions are hilar in distribution, tending to radiate outward along the bronchi; red hepatization is seen in the older areas. The lesions are rarely unilateral, not dense or homogeneous, and may produce a striated appearance.

Colonies that are hysterectomy-derived and barrier-maintained are free from PVM, indicating that the virus is not infective by transplacental routes. Eradication of PVM by use of SPF colonies is therefore practicable. In conventional colonies, the horizontal transmission of PVM should be susceptible to effective control by the use of filter-top cages.

Recommended Reading on PVM

Nelson, J. B., and J. W. Gowen. "The Establishment of a Rat Colony Free from Middle Ear Disease." *J. Exp. Med.* **54**: 629, 1931.

LACTIC DEHYDROGENASE VIRUS (LDH) INFECTION

This virus infection was described by RILEY et al. in 1960 as the cause of elevated lactic dehydrogenase activity in the plasma of tumor-carrying mice. It probably is endemic in wild mice. The infection is rare in laboratory mice and is yet another example of why wild mice should not be allowed access to feed, bedding, or mouse holding areas (RILEY et al., 1960).

In a more recent paper (1964), RILEY has noted that a synergism exists between this virus and the *Eperythrozoon coccoides* blood parasite. The hemolytic disease caused by *E. coccoides* is considerably more severe in the presence of LDH virus.

COXSACKIE VIRUS INFECTION

Coxsackie Group A viruses cause widespread lesions in skeletal muscle, whereas Group B viruses cause only minor lesions in skeletal muscle but do cause lesions in pancreas, brain, and liver. They can be differentiated by complement fixation tests (NIVEN, 1967). It is not known at the present how to treat or control this infection, and it is uncertain what the effect would be on most experimental procedures.

LYMPHOMAS

Of great interest to many now studying carcinogenesis and related therapy are the lymphomas of mice. They are of particular interest in animal-testing programs because of their incidence in certain inbred strains of mice, their probable viral etiology, and, most of all, their ability to appear in the middle of another, completely unrelated experiment. We have experienced this problem twice—once during a chronic murine pneumonia study in rats, and once in an immunology study in mice. SALAMAN (1967) has presented an excellent paper on a virus-induced lymphoma in mice, which gives much detail and is a must for anyone considering the lymphomas in mice— experimentally or diagnostically.

LEUKEMIALIKE DISEASE IN C57BL/6 MICE

RICHTER and BRICK (1967) have recently described a leukemialike disease in C57BL/6 mice. They observed that these mice, during a disease-control experiment, had a high incidence of a spontaneous leukemialike disease. Eleven of 28 (39%) 150- to 200-day-old mice had lesions in the lungs, liver, and/or kidney. Lung lesions were characterized by cuffs of lymphoctyes and reticulum cells around and within the walls of the pulmonary venules. Occasionally only one or two vessels were affected. In the liver, the lesions appeared as small scattered perivascular foci of mixed cell types, mostly lymphocytes and reticulum cells. In the kidney, similar foci were located around cortical arteries, or subjacent to the pelvic mucosa.

Initially, RICHTER and BRICK thought that the lung lesions were related to an infectious process; however, acute transmission experiments, employing tissue homogenates from affected mice, were unsuccessful. Their further observations on 300-day-old mice demonstrated marked progression in growth of lesions. At this time, the lesions were primarily reticulocytic, having lost most of their complement of lymphocytes, and many were then easily visible grossly on stained tissue sections of liver. Incidence at this time was 75% (15 to 20 mice). Distribution and morphological appearance resembled reticulum cell sarcoma. By day 400, there was little increase in lesion size, but eight of nine mice examined at this age were obviously affected.

SENDAI VIRUS INFECTION

Sendai virus is classified as a myxovirus, parainfluenza type I, and has also been called hemagglutinating virus of Japan (HJV), and influenza D virus. However, Sendai virus is the only member of the parainfluenza virus group known to naturally infect mice (TENNANT, 1966).

Natural infection of mice with Sendai virus has been reported in Japan, China, and Russia; only within the last few years has it been found to be a contaminant in breeder colonies in the United States. Sendai ranks third in prevalence of eight murine viruses that were tested: GDVII, Reovirus type 3, Sendai, PVM, MHV, polyoma, mouse adenovirus, and K virus. In addition, it is unusual in that the mean incidence of infection within infected colonies was 45%, whereas this value for almost all the other virus infections examined was approximately 20%, (N.B., GDVII is 38%) (PARKER et al., 1964; PARKER, 1966).

Sendai virus has been isolated from mice, rats, swine, and hamsters. It is closely related serologically to the human pathogen, hemadsorption virus 2; however, the possibility of human infection by Sendai virus has not been affirmed.

In Japan, Sendai virus appears in mouse stocks in the form of epizootics lasting 3 to 4 months. These infections are cyclic and the virus can be easily recovered from apparently healthy mice during the epizootics. The virus is actively spread among mice and cannot be isolated from them with antibody. In the United States, however, the infection in some colonies is of an enzootic type, and is continuous rather than cyclic. Infection of mice with Sendai virus occurs shortly after weaning, and is an acute and highly contagious enzootic infection contacting nearly all susceptible mice. In one study, approximately 45% of the mice sampled from positive colonies had demonstrable antibody.

Rats and guinea pigs reared in the same building with a positive mouse breeder colony have also been found to be positive for Sendai hemagglutination inhibition and complement fixation antibody. In addition, Sendai virus has been isolated from rats. Sera of 400 wild mice trapped in Maryland, Georgia, Florida, and Indiana were negative for hemagglutination inhibition antibody to Sendai virus. Factors accounting for the two patterns of infection, that is, enzootic and epizootic, probably reside in the ecological conditions prevailing in any individual colony.

Transmission of infection is direct through contact or indirect through fomite contamination. The target organs of Sendai infection in the mouse are the lungs, where it causes an interstitial pneumonitis.

High titers of virus are found in the lungs and saliva of infected mice. The isolation or detection of Sendai virus is readily accomplished in chicken eggs or cell cultures. Apparently the most sensitive method of detection is hemadsorption to infected monkey kidney cell cultures. Recovery from infection apparently provides life-long protection against reinfection.

MOUSE SALIVARY GLAND VIRUS (CYTOMEGALOVIRUS) INFECTION

The mouse salivary gland virus (MSGV) is almost identical in size and structure to the herpes virus and is thus classified in the herpes group (SMITH and RASMUSSEN, 1963). This virus has been detected in mice, rats, guinea pigs, hamsters, moles, monkeys, chimpanzees, and humans (BLACK et al., 1963; BRODSKY and ROWE, 1958; HARTLEY et al., 1957; MEDEARIS, 1964a,b; SMITH, 1959).

The MSG virus was initially grown on fibroblast cultures, for which it appears to be specific. The focal cytopathic effects induced by the guinea pig and human viruses are almost identical. On the other hand, the focal changes produced by the mouse virus are more numerous, less definitely outlined, and progress more rapidly in involving the entire culture.

The mouse virus apparently multiplies more readily than its guinea pig and human counterparts. Today, the MSG virus can be readily isolated in mouse embryo tissue cultures.

Both the intranuclear and cytoplasmic inclusions have been observed in the enlarged cells of the salivary glands in each species; however, the cytoplasmic inclusions are less common in the mouse. In experimental infections in the mouse and guinea pig, intranuclear inclusions of more recent infections are usually smaller than those occurring in infections of longer duration. In addition, in experimental infections, the cytoplasmic inclusions appear in the cells later than the intranuclear ones.

Utilizing tissue culture and suckling mouse procedures, urban and rural wild mice and colonies of laboratory mice were tested for the presence of virus in mouth swabs. The virus was detected in approximately 65% of the adult wild mice that were examined. In marked contrast, the virus does not seem to spread with facility in laboratory colonies, being present in only 2 or 3% of the mice (ROWE et al., 1962).

With only a few exceptions, the natural disease observed in animals has been of the latent type. The infection in inoculated and spontaneously infected mice takes the form of a prolonged low-grade infection of the salivary gland, with continual excretion of virus in the saliva for at least a year.

The phase of prolonged virus excretion is rarely accompanied by the presence of inclusion bodies, so virus isolation procedures must be employed. The virus can be isolated in mouse embryo tissue cultures, in which it produces a cytopathic effect. In addition, newborn mice provide a highly satisfactory virus isolation system, being about tenfold more sensitive than the tissue culture procedure.

The disease can be suspected in suckling mice by their characteristic appearance of malnutrition and a transverse striation pattern of the hair. The infection can be diagnosed with confidence by the gross appearance of the liver, which shows a pathognomonic type of yellow discoloration of the edges. There is no known treatment, but it does not appear in caesarean-derived colonies.

INFANTILE DIARRHEA

Two apparently distinct agents causing infantile diarrhea in mice have been described by KRAFT (1962). The agent called EDIM, epizootic diarrhea of infant mice, is apparently the cause of the classic disease described so often in the open literature, although other agents may also be involved. The second disease, called LIVIM, lethal intestinal virus of infant mice, bears a clinical resemblance to EDIM but can be separated clinically and pathologically from it. Both conditions present problems largely of diarrhea in young mice. Mice with EDIM exhibit lower mortality than LIVIM mice but do go through a period of clinical abnormal growth and appearance. The mice infected with LIVIM have a high mortality; an outbreak of this disease can completely ruin a breeding program, or production may drop for long periods of time to very low levels. At present, utilization of the filter-top cage, or one of its many modifications, a concept conceived by KRAFT (1958), will virtually eliminate from mouse colonies not only infant diarrhea but a host of other problems as well.

MOUSE POX (INFECTIOUS ECTROMELIA)

Mouse pox is caused by the murine representative of the pox group of viruses, *Poxvirus muris* (BURROWS, 1963; FLYNN, 1963; LANE-PETTER, 1963). This disease is one of the most important intercurrent infections in laboratory mouse colonies, outbreaks having been reported from many European countries and the United States. Colony infection can take the form of violent epidemic outbreaks or low-grade endemic conditions; or, it can often remain totally undetected until animals are subjected to some form of experimental stress.

The primary lesion develops at the portal of entry, but it is preceded by extensive multiplication of the virus and necrosis in the viscera, the spleen and liver particularly; acute cases show few specific lesions. The disease occurs in both a systemic (acute) form and a cutaneous (chronic) form. There may be some congestion of the liver and excess peritoneal fluid, but little else, whereas more chronic cases show rather more typical lesions. The liver may be pale, friable, and covered with grayish-white areas of focal necrosis. The spleen may be somewhat enlarged and similarly show necrotic areas. There may be some peritoneal exudate and often the contents of the small intestine in the duodenal region are blood-stained (this last lesion is sometimes lacking). Other organs in the abdomen and thorax are rarely involved.

Histologically, the liver in chronic cases contains numerous focal necrotic areas that may, in severe cases, coalesce. These lesions show no inflammatory infiltration. Necrosis of the spleen can also be very extensive; in mice that have recovered from severe chronic infections, such necrosed areas are replaced by fibrous tissue, a feature regarded as pathognomonic of this disease. Inclusion bodies can be demonstrated fairly readily in the epidermal cells of the skin in areas of inflammation. The epidermal cells are swollen and vacuolated, and many contain eosinophilic cytoplasmic inclusion bodies. Inclusion bodies can also be found in the acinar cells of the pancreas, and in the liver (but in the latter they are more difficult to demonstrate), which serves to differentiate other diseases with hepatic necrosis (e.g., salmonellosis).

Ectromelia is usually first suspected when mice bearing fairly extensive skin lesions are discovered. Earlier clinical signs are often missed. Ectromelia cannot be positively diagnosed clinically or by observing gross lesions, because the only consistent conditions—conjunctivitis, splenomegaly, and hepatic necrosis—are too nonspecific. Presumptive diagnosis depends upon the finding of suspicious skin lesions and the demonstration of a disease-producing agent in bacteriafree liver suspension. Use of the hemagglutination–inhibition test described by BRIODY (1959) will establish the diagnosis sufficiently for action to be taken in the event of an outbreak.

Clinical signs in acute cases are nonspecific. The animal becomes sick, lethargic, anoretic, emaciated, and develops a staring coat. The disease progresses rapidly, and death may occur in 24 hr. Inasmuch as the animals die before the dermal lesions appear, this form is not as contagious as the other. Infection is first indicated clinically by the presence of a primary lesion usually appearing as a swelling on the face. The disease may progress rapidly from this stage, and the animal may die with no more obvious clinical signs, or a generalized skin rash may develop over most of the body. This secondary rash in the chronic form consists of dry, scabby, sometimes very extensive lesions, which are particularly evident on the feet and tail, but which can be seen on the hair-covered parts of the body as well. Grossly, these lesions are indistinguishable from those produced by fighting or other forms of mechanical trauma. Occasionally the foot lesion progresses and the whole foot becomes necrotic and eventually sloughs off. Amputation is not an invariable or even frequent occurrence. Conjunctivitis occurs very frequently. This chronic form is considered highly contagious.

The chronic disease with skin and foot involvement can be readily induced experimentally by inoculation of liver or spleen suspension, containing virus, into the footpad of normal mice. Such animals show signs of sickness at about 7 days after inoculation, and by 13 or 14 days the typical skin rash has developed. Intraperitoneal inoculation of virus-containing material may result in acute or chronic disease, depending upon the dosage and virulence of the virus strains.

Outbreaks of the infection are likely to occur when fresh stock from other colonies is introduced into an animal house already containing infected mice. Outbreaks are also likely to be precipitated or detected among animals in use in the laboratory. The virus is often discovered during procedures involving tissue passage in mice, especially of tumors. A single positive diagnosis (one animal) is sufficient to indicate that a colony is infected and, unless control measures are instituted without delay, an epizootic is likely.

Stocks of animals bought at intervals should be quarantined for a period of 2 to 3 weeks before they are allowed to be introduced into the normal animal rooms. During this period, they should be carefully observed for signs of disease. Should a severe epidemic outbreak occur in a breeding colony, undoubtedly the safest procedure is to eliminate the entire stock, disinfect all the equipment and rooms with which these animals have been in contact, and refound the stock from known ectromeliafree sources. However, ectromelia can be effectively controlled by vaccinating all susceptible animals every 6 months. Sanitation and quarantine alone are ineffective.

The technique used in vaccinating is as follows. A microdrop of vaccine is placed about $\frac{1}{4}$ inch from the base of the tail on the dorsal surface. Seven or eight strokes are made through the microdrop with a 26-gauge needle. The skin should be scarified but frank hemorrhage should be avoided. To facilitate the scarification procedure, the point of the 26-gauge needle should be bent.

Any disease of mice characterized by ulcerating or scaling lesions of the head, tail, or extremities, and, even more important, any rapidly fatal disease of mice characterized by a high mortality and extensive necrosis of the liver should always be suspected of being ectromelia until proved otherwise.

The mortality rate is often high, but strains of the virus vary in this respect, and the infection may persist in latent form.

LYMPHOCYTIC CHORIOMENINGITIS

This disease is a widespread viral infection of laboratory mice which in its rare but most acute form produces a rapidly fatal paralyzing meningoencephalitis (BURROWS, 1963; LANE-PETTER, 1963; MAURER, 1964).

The virus occurs as a single antigenic type. Strains of LCM virus vary in their pathogenicity but there is no evidence of different antigenic types. The virus is filterable through the usual bacterial retaining filters, and is from 33 to 60 mμ in size. It is rapidly inactivated by ultraviolet light and is killed by 55C in 20 min. It has been cultivated in tissue cultures and chick embryos. The virus retains its viability when frozen or when dried and stored below 0C. Virus is present in all tissues and fluids of infected animals.

Lymphocytic choriomeningitis in mice is widespread throughout the United States, Europe, Asia, Africa, and probably the world. Surveys of wild house mice usually reveal about 10% to be infected. The disease has been found to occur naturally in mice, guinea pigs, chinchilla, cotton rats, foxes, dogs, monkeys, and man. Species that have been infected experimentally include rabbits, hamsters, squirrels, and horses. Infections are usually subclinical in the hamster, ferret, rabbit, dog, and horse. Cattle, pigs, cats, and chickens do not appear to be susceptible.

Transmission occurs readily by direct contact via the conjunctiva, respiratory or digestive tracts, or through the intact skin. The ease with which the virus may be transmitted to man has been amply illustrated by the many cases that have occurred in homes and laboratories. Numerous human cases have been reported in homes where infected mice were subsequently found. In most such cases, contaminated food or dust appears to have been the most likely source of infection.

Three forms of clinical infection occur with the disease: (1) an influenzalike illness with no other involvement; (2) the influenzalike infection, followed by meningeal symptoms; and (3) meningoencephalitis, sometimes fatal. When fatal, LCM usually produces a severe, nonsuppurative encephalomyelitis with lymphocytic infiltrations of the meninges, choroid plexus, and ependyma. There may also be perivascular infiltrations, engorgement, edema, focal gliosis, and necrosis.

Gross lesions that develop in mice, whether infected naturally or experimentally, are meager. There may be a pleural exudate, a fatty liver, and an enlarged spleen. Microscopically there is usually a round cell infiltration, predominantly lymphocytic, which is most striking in the meninges (especially at the base of the brain), in the choroid plexus, and in perivascular lymph spaces of submeningeal vessels. In general, the lymphocytic infiltrations in the brains of naturally infected cases are not as severe as in the experimental ones, which have been inoculated intracerebrally. The lungs frequently show peribronchiolar and perivascular infiltrations, with round cells and a slight thickening of alveolar walls. Small collections of lymphoid cells are frequently present in the liver near blood vessels and scattered through the parenchyma. There may be necrosis of hepatic cells in areas of lymphocytic infiltration. A patchy reticuloendothelial hyperplasia occurs in the spleen.

Diagnosis of LCM infection must rest upon (1) the isolation of virus from infected mice and cross-immunity tests with a known strain of the virus, (2) the demonstration of complement-fixing (CF) antibodies in mouse serum, and (3) the stimulation of inapparent or overt infection of intracerebral inoculation of sterile broth or starch suspension. Material from mice suspected of containing LCM virus can be inoculated intracerebrally into known uninfected mice or guinea pigs. Intracerebral inoculation in the guinea pig results in fever and death within 9 to 16 days.

There are three forms of infection in mice, depending upon the manner of infection—congenital, natural, or experimental. Mice infected *in utero*, or when inoculated during the first 7 or 8 days after birth, develop an infection that has several peculiar features. Most such mice appear normal yet harbor the virus as long as they live. They develop no circulating antibodies but are resistant to a clinical infection with lymphocytic choriomeningitis virus. The tissues, fluids, and exudates of such mice are highly infectious, and the virus readily spreads from them to other mice in the colony.

Natural infections that occur in mice after weaning differ from those which are congenitally infected in several respects. Naturally infected mice develop both complement-fixing and neutralizing antibodies. Their offspring are susceptible, and about 20% of them will show signs of clinical illness. Clinical illness usually occurs in mice under 6 weeks of age. With infected mice in a colony, usually over half the individuals in the colony will become infected. Clinical signs of infection, although nonpathognomic, are suggestive. Such mice appear rough, drowsy, and tend to sit in a corner by themselves. They become emaciated and develop photophobia and conjunctivitis. They become weak, but there is no paralysis. They tend to move only when pushed, and their movement is slow, stiff, and creeping. Because of the emaciation, their legs often appear too long. Any such suspicious animals should be removed from the colony; it is from such mice that virus can most often be isolated. Numerous factors influence LCM immunity in the mouse, making it an interesting immune mechanism for study.

A method for the recovery of virus from suspect mice is to remove the brain, spleen, and heart blood from a group of four suspects. These tissues may be weighed, pooled, and ground in a 20% concentration in buffered saline at pH 7.6. Centrifugation at 3000 rpm for 10 min will provide a sufficiently clear supernate for inoculation. The supernate should be cultured for bacterial contamination and, subsequently, 3- to 6-week-old mice may be inoculated intracerebrally with 0.03 ml and intraperitoneally with 0.25 ml. Because some strains of LCM virus are more readily isolated in guinea pigs than in mice, it is also well to inoculate young guinea pigs intraperitoneally with 0.5 ml of the suspect tissue supernate. The remaining inoculum can be frozen and, if later shown to be bacteriologically contaminated, penicillin and streptomycin may be added, followed by reinoculation of more animals. Inoculated mice should be held for 21 days. If lymphocytic choriomeningitis is present, symptoms will usually appear between the fifth and seventh day after inoculation.

Mice under 6 weeks of age, experimentally infected intracerebrally, become weak, tend to sit alone, have roughened fur, and frequently develop convulsions. These convulsions may be elicited in infected mice by twirling them by the tail. During a convulsive seizure, the hind legs are characteristically stretched out stiffly, the tail is rigid, the back humped, and the front legs move rapidly. They may drag themselves around for a few minutes by their front feet. They usually die in such a convulsive seizure with their hind legs extended in rigor. Death usually occurs in from 1 to 2 days after the onset of convulsions.

Naturally infected mice, which have only a subclinical infection, may be made to reveal this infection by the intracerebral injection of a noninfectious foreign material such as broth, serum, or a normal tissue suspension. The inoculation of foreign material into the brain tends to cause the virus to localize there, resulting in the same syndrome as if these mice had been inoculated intracerebrally with the virus. Such attempts to induce clinical infection should be done in mice under 6 weeks of age.

To develop and maintain a mouse colony free of LCM requires that disease-free breeding stock be obtained and kept isolated from wild rodents. The use of noncontaminated pelleted feeds and bedding, such as shavings from a source not exposed to wild rodents, is essential. Animal quarters must be free of insects and external parasites, which are potentially capable of transmitting the infection. Attendants must be alert to the fact that most species of laboratory animals are susceptible, that dogs and monkeys have been the source of infection in laboratories, and that mouse colonies should be kept separate from other species. Checks for latent virus should be made periodically, particularly in mice. If numerous intracerebral inoculations of control mice are made in the course of routine work, then latent natural infections would be detected.

Following the foundation of an LCM-free stock, some mice should be killed periodically and tested for the presence of virus. (If the animals are already being used for intracerebral inoculation work, this procedure would itself provide a continuing control.)

The infection of laboratory animals with LCM is obviously very undesirable. Its presence is likely to invalidate the results of experiments.

MOUSE ADENOVIRUS INFECTION

HARTLEY and ROWE (1960), during attempts to establish the Friend mouse leukemia virus in tissue culture, isolated a cytopathic agent possessing some properties of the adenovirus group. This agent was tentatively named "mouse adenovirus."

In suckling mice, the virus produces fatal disease with a pathological picture of inflammatory and necrotic foci in the heart and adrenals, and formation of acidophilic intranuclear inclusion bodies in these and many other tissues. In mice over 3 weeks of age, a nonfatal disease with mild myocarditis and focal necroses of the adrenals is seen.

Whether artificially or spontaneously exposed, infection in adult mice is prolonged. Excretion of large amounts of virus in the urine results in a rapid spread of the agent between cage mates. However, the spread throughout the room occurs at a much slower rate.

Of 34 colonies examined, 4 were positive for mouse adenovirus CF antibody (PARKER et al., 1966). The testing of sera from older mice—that is, those 4 to 5 months or more of age—may result in a higher incidence of infection with mouse adenovirus. As with many of the previous conditions discussed, hysterectomy-derived, barrier-maintained mouse colonies can remain virtually free of problems associated with these viruses.

RINGWORM (DERMATOMYCOSIS)

The dermatophytoses constitute a group of superficial fungus infections of keratinized tissues. They do not cause systemic infections and rarely invade subcutaneous tissues.

Dermatomycotic infections are common in both man and animals. They are generally known as "ringworm," owing to the fact that they typically begin in small areas and spread centrifugally, the region of greatest inflammation at the periphery forming an ever-widening circle. If the cases are not treated, new areas become infected and finally large areas of the skin may become involved through the coalescence of the primary lesions.

The fungi of the ringworm group grow almost exclusively on the keratinized layers of the skin, including the hair and other horny structures. They show a preference in most cases for the hairy portions of the body;

however, some species do occur on hairless areas. Typical infections usually involve the hair follicles and the hairs themselves. The latter become brittle and often break, or sometimes split. Affected hair becomes dry and lusterless, with the skin becoming scaly and harsh because of the formation of crusts. Bacterial complication is not uncommon, with pustules forming in the hair follicles. The secondary bacterial infections differ considerably in appearance, depending in part upon the nature of the infecting species.

Generally, ringworm thrives best in young animals, and in older ones that have been devitalized by disease or malnutrition. The infections occur most frequently on the face, neck, back, and tail; however, they may be found on any part of the body. Many types of ringworm are highly contagious, not only for members of the same species but often for members of other species also. Usually, when the infection occurs in an abnormal species, the disease does not spread in the new one but dies out in the initial host. Ringworm of most animal species has been known to infect man. Ringworm infections are extremely annoying and are often refractory to treatment. Some of the characteristics of the different genera follow.

Microsporum. Members of this genus attack only the hair and skin. The fungus is seen as a mosaic sheath of small spores surrounding the infected hair shaft. In the skin it appears only as segmented, branching mycelial elements, and cannot be distinguished from *Trichophyton* and *Epidermophyton.*

Trichophyton. Species of this group attack the hair, skin, and nails. Arthrospores may be arranged in parallel rows inside infected hair or on the external surface of the hair. In the skin and nails, species of *Trichophyton* appear as segmented, branching mycelial elements, which may or may not break up into arthrospores. Such forms are indistinguishable from those of *Microsporum* and *Epidermophyton.* This genus includes the causative agents of favus, a type of ringworm characterized by the production of structures called "scutula," which have the general appearance of shields. These lesions are formed as a consequence of the fungus radiating out from the hair follicles between the malpighian and the keratinized layers of the skin, a process that separates the latter from their attachments. The separated layers are thoroughly invaded by the fungus hypha, which form a felted mass within them. The hairs remain more or less intact, projecting out through the shields. The mycelium can be demonstrated inside the shafts of the plucked hairs as well as in the substance of the scutula.

Epidermophyton. This genus has only one species, and it attacks the hair and nails. In this type of infection, adnexae may be seen to contain branching mycelial elements that are identical in structure with the forms seen in *Microsporum* infections of the skin and *Trichophyton* infections of the skin or nails.

The dermatophytes have a worldwide distribution; however, some species are found constantly in certain geographic areas and rarely in others.

Lice and mites have been strongly suspected as being agents responsible for the transmission of fungal infections. Also, animal attendants have been known to have acquired the infection from animals, and possibly have aided its dissemination in a colony. The most common etiologic agent of ringworm in mice is *Trichophyton mentagrophytes*.

In severe cases, in addition to the typical appearance of affected skin and hair, there is a shiny appearance of the skin on the tail with a tendency toward obliteration of the ring pattern observable in normal mouse tails.

Trichophyton quinckeanum is an etiologic agent responsible for typical favus in mice. This fungal species is readily transmitted from mice to cats, in which lesions most commonly occur on the paws and ears.

Microsporum gypseum is an organism that causes ringworm in rats, mice, and wild rodents.

Immunity. Immunity in infected animals can be demonstrated not only by a cutaneous reaction to *Trichophyton*, but also by an immediate type of lesion produced on reinoculation. For varying periods of time after recovery from experimental infection, guinea pigs are immune to reinfection in the same site. There is only one effective systemic treatment, and that is with the antibiotic griseofulvin. Original experimental testing of this oral medication was performed on guinea pigs.

Control. It is doubtful that attempting to cure ringworm in laboratory mice is worth the effort. Topical applications are of doubtful value and oral treatments are quite expensive. Infected animals usually recover without treatment, but because they are likely to infect others before recovering it is probably most sensible to kill infected animals and effectively sterilize their cages.

Recommended Reading on Ringworm

Conant, Norman F. "Medical Mycology," in *Bacterial and Mycotic Infections of Man*, 4th ed., ed. by R. J. Dubos and J. G. Hirsch (Philadelphia: J. B. Lippincott Co., 1965), pp. 862–872.

Emmons, Chester W., Chapman H. Binford, and John P. Utz. *Medical Mycology* (Philadelphia: Lea & Febiger, 1963), pp. 86–119.

Hagen, William A. *The Infectious Diseases of Domestic Animals*, 5th ed., ed. by D. W. Bruner and J. H. Gillespie (Ithaca: Cornell University Press, 1961), pp. 512–521.

La Touche, C. J. "Ringworm in Laboratory Mice," in *The UFAW Handbook on the Care and Management of Laboratory Animals*, 2nd ed. (London: The Universities Federation for Animal Welfare, 1957), pp. 279–282.

Short, Douglas, and Dorothy P. Woodnott, eds. *The A.T.A. Manual of Laboratory Animal Practice and Techniques* (Springfield, Illinois: Charles C Thomas, Publisher, 1963), pp. 122–123, 139.

DEEP FUNGAL INFECTIONS

Although the secondary diseases of aspergillosis and penicilliosis appear to be common findings with no clinical or pathological sequelae, the more pathogenic fungal infections—such as actinomycosis, phycomycosis, cryptococcosis, coccidiomycosis, and sporotrichosis—are rare in laboratory mice. Diagnosis of these conditions is probably usually serendipidous unless some investigator happens to have this particular field of interest. Treatment would be impractical as most diagnoses are made histologically.

Recommended Reading on Deep Fungal Infections

Smith, John M. B., and Peter K. C. Austwick. "Fungal Diseases of Rats and Mice," in *Pathology of Laboratory Rats and Mice*, ed. by E. Cotchin and F. J. C. Roe (Philadelphia: F. A. Davis Company, 1967), pp. 681–732.

Recommended Reading on Infectious Mouse Disease

The following selected bibliography on infectious mouse diseases of possible research interest, either as models or as possible interfering mechanisms with experimental results, is listed alphabetically by author.

Balfour-Jones, S. E. B. "Bacillus Resembling *Erisipelithrix muriseptica* Isolated from Necrotic Lesions in Livers of Mice." *Brit. J. Exp. Pathol.* **16**: 236, 1935.

Barski, G. "Detection of Latent Ectromelia Virus Infection in Mice with the Aid of Human Cell Cultures *in vitro*." *Virology* **10**: 155–157, 1960.

Beverly, J. K. A. "Congenital Transmission of Toxoplasmosis Through Successive Generations of Mice." *Nature* **183**: 1348, 1959.

Bittner, J. J. "Virus Agent Associated with Mammary Gland Tumors in the Mouse." *Science* **84**: 162, 1936.

Braunsteiner, H., and C. Friend. "Viral Hepatitis Associated with Transplantable Mouse Leukemia. I. Acute Hepatic Manifestations Following Treatment with Urethane or Methylformamide." *J. Exp. Med.* **100**: 665, 1954.

Chang, Y. T. "Evolution of Murine Leprosy." *Am. Rev. Tuberc. Pulmonary Diseases* **79**: 805–809, 1959.

Culbertson, J. T. "The Elimination of the Tapeworm *Hymenolepsis fraterna* from Mice by the Administration of Atabrine." *J. Pharmacol. Exp. Therap.* **70**: 309–314, 1940.

DeBurgh, P., A. V. Jackson, and S. E. Williams. "Spontaneous Infection of Laboratory Mice with Psittacosis-like Organism." *Australian J. Exp. Biol. Med. Sci.* **23**: 107, 1945.

Findlay, G. M., E. Klieneberger, F. O. MacCallum, and R. D. MacKenzie. "Rolling Disease. New Syndrome in Mice Associated with PPLO." *Lancet* **235**: 1511, 1938.

Findlay, G. M., R. D. MacKenzie, F. O. MacCallum, and E. Klieneberger. "The Aetiology of Polyarthritis in Rats (PPLO)." *Lancet* **237**: 7, 1939.

Gard, S. "A Note on the Coccidium *Klossiella muris*." *Acta Pathol. Microbiol. Scand.* **22**: 427–434, 1945.

Giddens, W. E., K. K. Keahey, G. R. Carter, and C. K. Whitehair. "Pneumonia in Rats Due to Infection with *Corynebacterium kutscheri*." *Pathol. Vet.* **5**: 227–237, 1968.

Kendall, A. I. "New Species of Trypanosome Occurring in Mouse (*Mus musculus*)." *J. Infect. Diseases* **3**: 228, 1906.

Krakower, C., and L. M. Gonzales. "Mouse Leprosy." *Arch. Pathol.* **30**: 308–329, 1940.

Li, C. P., and W. G. Jahnes. "Hydrocephalus in Suckling Mice Inoculated with SE Polyoma Virus." *Virology* **9**: 489–492, 1959.

Marmorston, J. "Effect of Splenectomy on a Latent Infection, Eperythrozoon Coccoides in White Mice." *J. Infect. Diseases* **56**: 142–152, 1935.

Nelson, J. B. "Association of a Special Strain of PPLO with Conjunctivitis in a Mouse Colony." *J. Exp. Med.* **91**: 309–320, 1950.

———. "The Selective Localization of Murine PPLO in the Female Genital Tract on Intraperitoneal Injection in Mice." *J. Exp. Med.* **100**: 311–320, 1954.

———. "Emergence of Hepatitis Virus in Mice Injected with Ascites Tumor." *Proc. Soc. Exp. Biol. Med.* **102**: 357–360, 1959.

Rodaniche, E. "Sulfanilylguanidine and Sulfanilamide in the Treatment of Lymphogranuloma Venerum Infections of Mice." *J. Infect. Diseases* **70**: 58, 1942.

Shechmeister, I. L. "Pseudotuberculosis in Experimental Animals." *Science* **123**: 463, 1956.

Smith, T., and H. P. Johnson. "On a Coccidium Parasite in the Renal Epithelium of the Mouse." *J. Exp. Med.* **6**: 303–316, 1902.

Thompson, J. "Salivary Gland Disease of Mice." *J. Infect. Diseases* **58**: 59, 1936.

Tyzzer, E. E. " 'Interference' in Mixed Infections of Bartonella and Eperythrozoon in Mice." *Am. J. Pathol.* **17**: 141–153, 1941.

Wenrich, D. H. "Trichomonad Flagellates in the Coecum of Rats and Mice." *Anat. Record* **29**: 118, 1924.

Winter, W. D., Jr., and G. E. Foley. "Enhancement of Candida Infection: Differential of Renal Lesions in Mice Treated with Aureomycin." *J. Infect. Diseases* **98**: 150, 1956.

Parasites Infecting Mice

INTESTINAL PARASITES

Nematodes. Heterakis spumosa is found in the large intestine and caecum of wild rats but apparently is only experimentally infectious to mice. At least, it is unimportant to laboratory mice.

Aspicularis tetraptera is an oxyurid nematode or pinworm. It is fairly prevalent in conventional mouse colonies and can be diagnosed by putting fecal material from the caecum or colon into tap water under a dissection scope and observing the adult worms. We usually differentiate it from *Syphacia obvelata* by the appearance of the eggs in the gravid female, or free in the water-fecal mixture. The egg of *Aspicularis* is symmetrical, whereas the egg of *Syphacia* is asymmetrical and almost crescent shaped. This parasite can be adequately treated with 2,2,2-trichloro-1-hydroxyethyl phosphonate and atropine 1:180 in the same method as that described for *Syphacia*. This parasite does not cause a serious disease and is not very infectious in our experience. Although it has a direct life cycle, it does not appear to spread rapidly within a colony.

Syphacia obvelata is an oxyurid nematode that is practically endemic in most conventional mouse colonies. It is a pinworm similar to *A. tetraptera* in that it looks very much like it on gross observation. *Syphacia* has a direct life cycle. The simplest method of diagnosing *Syphacia* is to Scotch-tape the perianal area and then to apply the tape to a slide and observe under a microscope at 10 × magnification. The egg will be found on the slide, usually in an area of hair that has adhered to the tape. The gravid female apparently lays her eggs on the hairs of the perineal region. This parasite is one of the most infectious and troublesome parasites in a mouse colony. It is so infectious from room to room, and even from animal facility to facility, that it can be used to advantage as an indicator organism in a pathogenfree animal facility. It can be transmitted by caretakers and one often gets the impression that it may float through the air. The most efficient treatment we have found is an organic phosphate called Trichlorfon (Fort Dodge Laboratories), which is 0,0-dimethyl 2,2,2-trichloro-1-hydroxyethyl phosphonate combined with atropine 1:180. It is dissolved in the drinking water with sugar to mask the bitter taste. After 2 weeks of treatment the pinworms are completely eliminated (SIMMONS et al., 1965).

Nippostrongylus muris is a nematode that is found in the upper intestinal tract of the mouse. The infection may occur by larval ingestion or skin penetration with a larval migration through the lungs. Heavily infected animals may show marked lung pathology. With adequate sanitation procedures, this parasite is not a usual pathogen in mouse colonies.

Cestodes. *Hymenolepsis nana*, the so-called "armed dwarf" tapeworm, is found in the small intestine. The important aspects of infection with this parasite are its infectivity to humans and its direct life cycle. It can reinfect its host within about 1 month. However, in mouse colonies, where bedding is frequently changed and sanitation is held at a high level, this parasite is rare and practically self-eliminating.

Hymenolepsis diminuta is an unarmed tapeworm requiring an intermediate host of some sort. It is found in the ileum. Beetles and fleas are common intermediate hosts and then become infective. This tapeworm is not apparently infective to man inasmuch as humans do not ordinarily ingest fleas and beetles. (Man does, however, ingest a certain amount of mouse fecal material and can become infected with *H. nana*.) It is almost nonexistent in well-run mouse colonies because of the need for intermediate hosts. If proper sanitation and pest control are observed, *H. diminuta* will be self-eliminating. This parasite is not a very important mouse pathogen because it occurs only sporadically, and even then conditions must be extremely fortuitous.

Coenurus serialis is a larval stage of *Taenia serialis*, a dog tapeworm. It is not an important parasite of laboratory mice.

Oochoristica ratti is a rare tapeworm of little importance to laboratory mice.

Cysticercus fasciolarus is a bladder worm, the larval stage of *Taenia taeniaformus*, which is found in the cat as an adult tapeworm. The cysticercus may appear in the liver of mice if they have for some reason been exposed to cat fecal material. It is very important to keep cats out of mouse-holding areas, as well as feed and bedding storage areas.

Protozoa. Entamoeba muris, Trichomonas muris, Chilomastix bettencourti, Giardia muris, Hexamita muris, Trypanosoma duttoni, Cryptospiridin muris, and *Hepatozoon muris, Toxoplasma gondii, Klossniella muris,* as well as several species of *Eimeria,* have been reported in mice. In most protozoan diseases of mice, it is clearly undesirable to have the infections because they may be indicators of poor sanitation conditions and external parasite infestations. In short, they may act as indicator systems for all ecological classes of mice. They may be important if they are discovered in shipments of animals from commercial sources.

Acanthocephala. Acanthocephala or thorny-headed worm, *Moniliformis moniliformis,* is a mouse parasite requiring an insect intermediate host such as a cockroach. It is yet another example of the necessity of keeping mouse colonies free from pests of all kinds. Those mice that ingest infected insects will become hosts and will pass embryonated eggs that may then be again ingested by the intermediate hosts. It is unlikely that this parasite could be an important pathogen in a properly managed mouse colony.

EXTERNAL PARASITES

Mites. Psorergates simplex is a parasite of mice that appears as a whitish nodule on the visceral surface of the skin. The mites in all stages of development are found in subdermal cysts (possibly hair follicles). It is not a serious problem in laboratory mice, but may go unnoticed in many instances. It may also occur as an ear mange mite, in which instance it can be treated with Canolene or some other dog ear mite preparation.

Myobia musculi, Ornithonyssuc bacoti, Radfordia affinis, Myocoptes musculinus, Myocoptes romboutsi, and *Notoedres muris* are all mites that may be identified on laboratory mice. *Polyplax serrata* (a louse) and various species of fleas may also be found on mice. The species differentiation is of little importance except for academic interest. Animals infected with external parasites are more irritable and usually bite and scratch, causing skin lesions that become secondarily infected. *Polyplax serrata* is involved in transmission of *Eperythrozoon coccoides*. The best method of eliminating external parasites is through rederiving the strain and housing them in filter-top caging or by acquiring new animals from parasitefree stocks and then housing them in filter-cap cages. We have eliminated a wide variety of external parasites with whole-body immersion in 2% Malathion preparations, using two dippings at 2-week intervals. This treatment is rather drastic, with approximately 3 to 5% mortality; however, it is highly effective if the animals can be moved to a clean environment. Many other powders and the like may be used, but they generally only decrease rather than eliminate the external parasite problem.

BLOOD PARASITES

We have elected to divide the blood parasites found in mice into two groups—namely, protozoa and rickettsia.

Protozoa—such as *Plasmodium berghei, P. vinckei, Toxoplasma gondii, Trypanosoma leivisi, T. cruzi, Hepatozoon muris,* and *Babesia (Piroplasma) muris*—have been found in various cells in the blood stream. Routine screening of blood smears for these organisms in this laboratory have produced only negative results to date.

Rickettsia. Eperythrozoon coccoides is a blood parasite that belongs with the order Rickettsiales, although it has been classified as a protozoon in the past. The organism may persist for long periods in the latent state only to be activated by some stress condition, such as X irradiation, splenectomy, or concurrent disease.

When the latent disease is exacerbated by a stress factor and becomes active, a mild-to-severe hemolytic anemia and corresponding reticulocytosis occur. The disease is not usually fatal but it may be if the stress is severe enough.

Diagnosis of latent infections is based on splenectomy of the mice and examining blood smears 3 days postsurgically, and every 2 to 3 days thereafter up to day 10, until the organism appears on erythrocytes. If still negative after 2 to 3 weeks, one may be reasonably sure that it is not present. It is a difficult diagnosis to make because so many artifacts may resemble the organisms on the erythrocytes.

Animal treatment has included tetracyclines and arsenicals, but caesarean derivation is probably the best system. There is always a danger in introducing tumor stocks and other body fluids into mice from other sources because the organism may have been latent in the donor animal. In conventional colonies, the main control method should be control of external bloodsucking arthropods such as *Polyplax spinulosa*. This technique has been studied in the *Hemobartonella muris* infection in rats (CRYSTAL, 1959).

BAKER and LINDSEY (1967) have written a very comprehensive discussion on latent Bartonellosis as a complicating factor in animal experimentation. We would like to quote directly from their paper three excellent examples of the synergistic behavior of *E. coccoides*.

(1) Mouse Hepatitis Virus (MHV)—During studies of mouse hepatitis virus Gledhill and others[44,56] noted strain variations in susceptibility of mice to experimental infection. These workers successfully separated from their inocula two components, neither of which produced overt disease alone, whereas their combined effect was fatal hepatic necrosis. Both components were transmissible and filterable; however, one was found to be inactivated

by exposure to temperatures of 38°C for 24 hr and sensitive to the *in vivo* effects of aureomycin and terramycin. This labile component of Gledhill's filtrates has been identified as *E. coccoides*.[33,44] *Eperythrozoon* infections apparently potentiate the virulence of MHV. Furthermore, variations in susceptibility between mouse strains reflect, at least in part, the presence or absence of latent *E. coccoides* in experimental mouse colonies.

(2) Lactic Dehydrogenase Virus (LDH or Riley agent)—A virus causing a five- to tenfold elevation in plasma lactate dehydrogenase enzyme concentration has been recovered from tumor-bearing mice. The original characterization of this agent included its implication in the induction of anemia in inoculated mice.[54] Stansly and his collaborators[41] observed that an organism intimately associated with the LDH agent caused a threefold increase in spleen weight and named this agent "The Transmissible Spleen Weight Increase Factor" (SWIF). Riley and others recently discovered the almost universal contamination of LDH agent with *E. coccoides*.[54,57] Stansly and Neilson[43] concluded that the agent they originally described as SWIF was in fact *E. coccoides*. LDH agent and *E. coccoides* have been successfully separated and the specific effects of each entity evaluated.[54] These studies have revealed that *E. coccoides* is primarily responsible for the anemia and splenomegaly, whereas the enzyme elevating effect is the major response to the LDH agent. It is of interest to note that in combination the organisms act synergistically to enhance the elevation of serum LDH concentrations.

(3) *Plasmodium sp.*—The influence of intercurrent bartonella infections in altering the pattern of malarial infections and host response has been adequately demonstrated.[47,58] Concurrent hemobartonellosis or eperythrozoonosis reduce the pathogenicity of malarial infections, shorten the prepatent period and result in anemias inconsistent with the degree of plasmodial parasitemia. The observations of Ott and Stauber[48] are especially significant. These workers noted that in the presence of *E. coccoides*, *Plasmodium chabaudi* infection in mice was the low-level, chronic, nonlethal disease usually ascribed to this malarial parasite. When, however, the *Eperythrozoon* were eliminated, the *P. chabaudi* infection assumed a fulminating, fatal infection characteristic of *P. vinckei*. Other data have been presented to support the presumption that these malarial species may indeed represent the same organism.[1]

OTHER PARASITES (BY ORGAN LOCATION)

Sarcocystis. *Sarcocystis muris* is a protozoan parasite in muscle tissue. It is of little importance to mouse colonies except as an incidental finding at autopsy. This parasite shows up on hematoxylin-eosin stained muscle slides as a blue encapsulated body. There may be one or many in one

---

[1] Literature cited in the quoted material, as indicated by the superscript numbers, may be identified by referring to the quoted source, BAKER and LINDSEY (1967).

section. No therapy is known, but the parasite is not seen in SPF colonies; therefore, it can presumably be eliminated by derivation techniques.

Encephalitozoonosis (Nosematosis). *Encephalitozoon cuniculi*, a protozoan organism recently renamed *Nosema cuniculi* (LAINSON et al., 1964), has been implicated as a cause of disease for many years, perhaps causing much more disease than we are aware. The organism occurs in groups or clumps surrounded by an envelope-type cell membrane. It is seen in epithelial cells, urine, brain tissue, and numerous other places. The life history is incomplete at present. The clinical disease is poorly defined and vague to all but the experimentalists who study the organism. It is probably usually latent unless an experimental stress occurs. Encephalitis and febrile response may result. This disease is probably misdiagnosed unless careful histology is done. No treatment is known.

The larval forms of several internal parasites of other animals as well as of mice may be encountered during postmortem and microscopic examination for other reasons. The following is a list of parasites and/or larval forms and the most likely area in which to find them.

Parasite	*Location*
Taenia pisiformis	liver
Taenia hydatigena (larva of)	mesenteries
Multiceps serialis (larva of)	connective tissue
Trichinella spiralis (larva of)	muscle
Capillaria hepatica	liver
Hepaticola hepaticola	liver
Trichosomoides crassicuda	urinary bladder

Recommended Reading on Parasites

Habermann, R. T., and Fletcher P. Williams, Jr. "The Identification and Control of Helminths in Laboratory Animals." *J. Natl. Cancer Inst.* **20**: 979–1009, 1958.

Keefe, Thomas J., J. E. Scanlon, and Lenly D. Wetherald. "*Ornithonyssus bacoti* (Hirst) Infestation in Mouse and Hamster Colonies." *Lab. Animal Care* **14**: 366–369, 1964.

Noble, E. R., and C. A. Noble. *Animal Parasitology, A Laboratory Manual* (Philadelphia: Lea & Febiger, 1962).

Oldham, J. N. "Helminths, Ectoparasites and Protozoa in Rats and Mice," in *Pathology of Laboratory Rats and Mice*, ed. by E. Cotchin and F. J. C. Roe (Philadelphia: F. A. Davis Company, 1967), Chap. 20, pp. 641–679.

Experimental Disease

It is impossible to discuss or even list all the experimental diseases that have been studied in the mouse. Many millions of mice have been used in studying infectious, noninfectious, nutritional, and genetic diseases. Natural and artificial disease models have evolved until today there is probably no classification of mouse disease, however exotic, that is not being studied intensively. In this section, we mention a few of the most interesting experimentally induced diseases.

MYCOBACTERIUM LEPRAE

Mycobacterium leprae has been studied in mice since it was found that thymectomized-irradiated mice are good experimental subjects for this disease organism (REES, 1966; REES *et al.*, 1967). According to recent work on this organism in mice (SHEPARD and CONGDON, 1968), the histopathological studies showed that in untreated controls the organism was not found out of the immediate area of the injection site in the foot pad. However, in mice infected in the same location with the same numbers of the organism, but treated by thymectomy and with X irradiation and syngeneic bone marrow, the bacilli were found in regional lymph nodes, in other feet, and even in ears and nose. It is a very interesting and ingenious method of manipulating an aberrant host animal to meet the experimental requirements. This sort of combination of surgical, immunosuppressant adaptation will probably become more common in future studies of difficult and fastidious organisms.

SECONDARY DISEASE

Radiation chimera, as the term is used by most investigators, refers to the lethally irradiated animal that has been kept alive by transplantation of hematopoietic cells after an acute radiation exposure.

In the foreign bone marrow chimera, however, a secondary disease, frequently fatal, develops in the early weeks after irradiation, even though the initial transplantation was entirely successful (CONGDON and URSO, 1957; DE VRIES and VOS, 1959). The disease and death cannot be attributed to marrow failure because adequate amounts of the bone marrow transplant are usually present at autopsy.

From studies carried out during the past decade, it is not the consensus that immunologically competent cells from the foreign transplant attack the recipient in a graft-versus-host reaction to initiate the secondary disease process (TRENTIN, 1958).

The manipulations of whole-body systems represented by the establishment of this disease in mice, and all the subsequent methods of treating it, make it one of the most fascinating of the experimentally induced diseases (CONGDON et al., 1965, 1967). Although it is not specific for mice, the numbers needed to statistically interpret results usually mean that large numbers of mice are used in the study of the secondary disease syndrome.

Recommended Reading on Experimental Disease

Other interesting experimentally induced diseases of mice that have added to the massive backlog of information about the use of the mouse over a period of many years are described in the literature listed in the following bibliography.

Beswick, T. S. L. "Further Observations on Experimental Herpes Simplex Infection in the Baby Mouse." *J. Pathol. Bacteriol.* **79**: 69–76, 1960.

Geiger, J. C., and K. F. Meyer. "Experimental Food Poisoning in White Mice with Heat Stabile Paratyphoid Poisons." *Proc. Soc. Exp. Biol. Med.* **26**: 91, 1928.

Innes, J. R. M., E. J. Donati, and P. P. Yevich. "Pulmonary Lesions in Mice Due to Fragments of Hair, Epidermic and Extraneous Matter Accidentally Injected in Toxicity Experiments." *Am. J. Pathol.* **34**: 161, 1958.

Kalter, H. "Congenital Malformations of the Rectum and Urogenital System Induced by Maternal Hypervitaminosis A in Strains of Inbred Mice." *Anat. Record* **136**: 219, 1960.

Larsh, J. E., and G. J. Rase. "A Histopathologic Study of the Anterior Small Intestine of Immunized and Nonimmunized Mice Infected with *Trichinella spiralis*." *J. Infect. Diseases* **94**: 262, 1954.

Perrin, T. L., and I. A. Bengston. "The Histopathology of Experimental 'Q' Fever in Mice." *Am. J. Pathol.* **34**: 161, 1958.

Skinner, H. H. "Propagation of Strains of Foot and Mouth Disease Virus in Unweaned White Mice." *Proc. Roy. Soc. Med.* **44**: 1041–1044, 1951.

Sokoloff, L., O. Mickelsen, E. Silverstein, G. E. Jay, Jr., and R. S. Yamamoto. "Experimental Obesity and Osteoarthritis." *Am. J. Physiol.* **198**: 765–770, 1960.

Straub, M. "The Microscopical Changes in the Lungs of Mice Infected with Influenza Virus." *J. Pathol. Bacteriol.* **45**: 75–78, 1937.

Genetic Diseases (Mutants and Hereditary Abnormalities)

Inasmuch as many papers, seminars, and textbooks have been devoted to genetic conditions occurring in mice, we will include here only a reference list for some of the reported abnormalities. This list will include several reports published by the Jackson Laboratory (Bar Harbor, Maine) because many of the detailed genetic studies are conducted there.

Recommended Reading on Genetic Disease

Bernstein, S. E., and L. C. Stevens. "Jaundice, a Congenital Microcytic Anemia of the House Mouse." *Proc. Tenth Intern. Congr. Genet.* **2**: 25, 1958.

Brat, V., R. Shull, R. B. Alfin-Slater, and B. H. Ershoff. "Occurrence of Osteoporosis in Mice with Muscular Dystrophy." *Arch. Pathol.* **69**: 649–653, 1960.

Center, E. M. " 'Dorsal Excrescences' in the Mouse—Genetic or Non-genetic?" *J. Heredity* **51**: 21–26, 1960.

Davis, T. R. A., and J. Mayer. "Imperfect Homeothermia in the Hereditary Obese Hyperglycemic Syndrome of Mice." *Am. J. Physiol.* **177**: 222–226, 1954.

Gowan, J. W. "Mouse Inheritance as It Contributes to Disease." *Carworth Quart. Letter* No. 60, 1961.

Green, Earl L., ed. *Biology of the Laboratory Mouse*, 2nd ed. (New York: McGraw-Hill Book Company, 1966), 706 pp.

Green, Margaret C. (a) *Linkage Map of the Mouse* (Bar Harbor, Maine: The Jackson Laboratory, June 1965). (b) *Neurological Mutants of the Mouse* (Bar Harbor, Maine: The Jackson Laboratory, June 1965), Tables 1 and 2.

Lane, Priscilla W. *Lists of Mutant Genes and Mutant-Bearing Stocks of the Mouse* (Bar Harbor, Maine: The Jackson Laboratory, June 1966).

Pappenheimer, A. M., F. S. Cheever, and H. Salk. "Jaundice in Mice Due to Anomalies of the Biliary Tract." *J. Exp. Med.* **101**: 119, 1955.

Silverstein, E. "Primary Polydipsia and Hydronephrosis in Inbred Strains of Mice." *Acta Endocrinol. Suppl.* **51**: 121, 1960.

Wegelius, O. "The Dwarf Mouse—An Animal with Secondary Myxedema." *Proc. Soc. Exp. Biol. Med.* **101**: 225, 1959.

Neoplastic Disease

Many papers, and even books, have been devoted to the subject of neoplasia. Because of its predictable pattern of tumor development, rapid reproduction, and short life-span, the mouse has been utilized for a major portion of the tumor research done thus far. It is not our intent here to enter into a detailed discussion concerning definition or classification of mouse neoplasms, nor to examine deeply the many types of tumors that may be experimentally induced. However, selected tumors that rarely occur or that occur in low numbers but are augmented by certain experimental procedures will be mentioned briefly, along with specific references. We also describe strain-related spontaneous tumors occurring in mice. A better choice of experimental animal, strain, or sex may be determined if one knows in advance the expected incidence of a particular neoplasm. Known virus-induced neoplasms—such as polyoma, leukemia, and so on—were covered earlier in the virus section of this chapter.

Spontaneous tumors in mice are found in almost every site in the body. Certain inbred strains of mice have a very high frequency of mammary tumors and lymphoid leukemia. Other tumors—such as lung, liver, and ovarian tumors—occur at somewhat lower incidence, although certain strains may have a very high incidence of a specific type of neoplasm; for example, the A and SWR strains are reported to be highly susceptible to lung neoplasms (HESTON, 1966). By far the majority of tumors arise in older mice—that is, animals over 10 months of age—although mammary tumors, leukemia, and adrenal cortical tumors arise between 6 and 10 months in strains such as C3H, DD, AKR, and CE.

The following list of inbred strains and commonly occurring tumors (Table 3.3) may be found in greater detail in GREEN (1966, Chap. 27, by Edwin Murphy).

TABLE 3.3

Incidence of Common Tumors in Various Inbred Strains of the Laboratory Mouse

Tumor type	Mouse strain	Incidence	Age, sex, condition of mice
Mammary gland tumors	C3H/An	High (99–100%)	Both breeding and virgin females
	C3HeB	49%	Late-occurring breast tumors
	DBA/2	25–30%	Higher in breeding mice than in virgins
	C3H/He	Up to 100%	Breeding females
	A	High	Breeding females
	C3H/Bi	High	Breeding females
	DD	75–80%	About the same in breeders and virgins
	CBA	Wide range (5–65%)	Breeding females
	CBA	75%	Virgin females
	DBA	High	Breeding females
Lymphocytic leukemia	AKR	Over 90%	Either sex
	C3H/Fg	96%	Virgin females

TABLE 3.3 continued

Lymphocytic leukemia continued	C3H/Fg	89%	Males
	C58	Very high	Either sex
	SJL	75%	Females
Primary lung tumors	NH	30–42%	Older mice
	A	Very high (90%)	Either sex
	CC57W	25%	Both sexes
	SWR	80%	Either sex
	BALB/c	2.5%	
(adenomas)	NZO	High	
Hepatoma	C57BL	25%	Males
	C3HeB	Very high	Older breeding males
	C3HeB	58%	Virgin females (lower in breeding females)
	C3H/He	85%	Males
	CBA	64%	Males
Reticulum cell neoplasm, type B	SJL	Very high (80%)	Virgin or breeding females; breeding males
Reticulum cell neoplasm, type A	C57BL	15%	Breeding females
Ovarian tumor	C3HeB/De	From 37% up	Breeding females
	C3HeB/De	To 47%	Virgins
	C3HeB/Fe	Low (22%)	Virgin females
	C3HeB/Fe	Moderate (64%)	Breeding females
	R III	50–60%	Breeding and virgin females
	CE	34%	All females
Pituitary tumors	C57L	33%	Breeding females
	C57BR/Cd	33%	Breeding females
Hemangio-endothelioma	HR/De	24%	Males and females
Adrenal cortical tumor	CE	100%	Ovariectomized females
	CE	79%	Castrated males
Myoepithelioma	BALB/c	4%	Either sex
	A	4%	Either sex
Skin papilloma	HR/De	9%	Hairless mice
Skin carcinoma	HR/De	3%	Hairless mice
Subcutaneous sarcoma	C3H/J	3%	Females

Nutritional Deficiencies

Many specific pathologic effects of nutritional deficiencies may go unnoticed unless severe lesions develop. Many times a decreased growth rate or altered reproduction performance may be the only noticeable signs of a nutritional deficiency. Specific diseases normally result from a deficiency of a required nutrient, although nutritional imbalance, reduced intestinal synthesis, poor absorption, or improper utilization of absorbed nutrients may also result in obscure symptoms.

Conditions related to stress factors (such as overcrowding, caging, artificial light, improper air changes, extreme variation in temperature and/or humidity, and frequent handling) can enhance the effects of improper diet or dietary deficiencies. Animals in a good state of nutrition are usually healthy and free of symptomatic disease. Malnutrition is known to cause a decreased resistance to many bacterial and rickettsial infections, and can aggravate parasitic infection and precipitate protozoal infections. Vitamin deficiencies may result in a poor antibody response as well as reduced phagocytic activity, altered tissue integrity, interference with normal tissue repair and replacement, interference with the production of nonspecific protective substances, and interference with the destruction of bacterial toxins. Indeed, alterations may occur in the intestinal flora, and the endocrine and electrolyte balances are affected by disrupted nutrition.

PROTEIN

Owing to the major concern of maintaining a high production performance from breeding colonies, diet plays a very important role. Although many specific nutrients are blamed for poor production, a protein deficiency is a very logical place to begin to solve this problem. Growth of young mice may be severely inhibited, whereas mature mice usually show rapid weight loss in the form of muscular wasting. Death of either young or old mice may occur in prolonged deficiencies. Specific problems associated directly with breeding mice include interruption of normal estrus, birth of weak or dead offspring, and fetal resorption. Protein-deficient diets also adversely affect normal lactation in breeder females.

FATS

Lactating females have a greater requirement for fats than do non-lactating mice. HOAG et al. (1966) reported decreased productivity from DBA/2J mice by changing the diet from 6 to 11% fat, and changing from saturated fats to unsaturated fatty acids. Fat-deficient diets are more commonly incriminated in cases of dry hair, dry scaly skin, and increased dermatitis. In an experiment in which mice received a diet enriched with animal fat (25% lard), accelerated articular aging and accentuated evolution of osteoarthritis were observed. Increasing the fat level of the diet with vegetable fat (25% commercial cottonseed oil) did not affect the mice to the extent that was seen with animal fat (SILBERBERG and SILBERBERG, 1960). As reported by MORRIS (1944), the mouse is unable to grow normally and maintain normal health in the absence of unsaturated fatty acids of the linoleic or linolenic acid series.

CARBOHYDRATES

Carbohydrates apparently have caused very little problem as a result of deficiencies in mouse diets. Specific deficiency symptoms in mice fed carbohydrate-free diets have not been reported. Certain other laboratory animals require the presence of bulk, such as natural food fiber or cellulose, but mice are not affected by its absence.

MINERALS

Much of the work reported concerning mineral deficiencies relates to more animals than just the mouse. Very little research has been devoted to a specific mineral deficiency in mouse diets. Imperfect calcification of epiphyseal plates, generalized body stiffness, joint enlargement, tetany, and increased levels of serum alkaline phosphatase have been reported as symptoms of calcium and phosphorous deficiency or imbalance (LOOSLI, 1963). It has also been reported (BEARD and POMERENE, 1929) that mice are susceptible to the same type of rickets found in rats, associated with a vitamin D deficiency and disproportion between calcium and phosphorus in the diet.

Iron is required for normal reproduction and growth of the offspring. Decreased litter size and birth weights are occasionally seen as a result of iron deficiency. Iron-deficient offspring show characteristic anemia symptoms.

A manganese deficiency has been suggested as a cause of poor growth of offspring, weak or dead offspring, and infertility in breeders. No reproduction in mice on low-manganese diets has been reported by SHILS and McCOLLUM (1943). It is generally agreed that mice, like rats, require manganese for successful lactation. A recent publication (ERWAY et al., 1966) links a manganese deficiency during prenatal development in normal mice and the presence of mutant genes to a specific congenital ataxia resulting from defective development of the otoliths.

Zinc has been reported to be necessary in the mouse diet. Deficiencies, as reported by DAY and SKIDMORE (1947), include impairment of growth, weakness, emaciation, greasy fur, hair loss, and death.

More is known about signs observed in mice fed potassium-deficient diets than many of the other minerals. Dry-hair coat, short broken hairs, and scaly tails have been associated with potassium deficiency. General emaciation and eventual death were also observed and reported by BELL and ERFLE (1958).

Vitamin studies, as related to mouse diets, have received much more extensive research in the area of required amounts and deficiency symptoms. Vitamin D deficiency was mentioned in its relationship to calcium and phosphorus, and therefore will not be listed again with the rest of the fat- and water-soluble vitamins.

FAT-SOLUBLE VITAMINS

Vitamin A requirements for the mouse have been reported by many investigators (WOLFE and SLATER, 1931; MCCARTHY and CERECEDO, 1952; MORRIS, 1947). Vitamin A deficiency symptoms are numerous and may be observed in relation to other specific vitamin deficiencies. The symptoms include squamous metaplasia of epithelial tissue, corneal opacity, eye exudates, incoordination, tremors, nerve degeneration, poor growth, rough-hair coat, rectal and vaginal hemorrhages, abortion, resorption, and irreversible sterility in males.

Hypervitaminosis A can also be a problem, and malformations of the mouse fetus owing to excessive amounts of vitamin A have been produced experimentally.

Symptoms of Vitamin E deficiency have been studied and/or reported by BRYAN and MASON (1940), GOETTSCH (1942), PAPPENHEIMER (1942), CERECEDO and VINSON (1944), MORRIS (1944), and LOOSLI (1963). Deficiency symptoms include muscular dystrophy and hyaline degeneration, weakness and incoordination, steatosis, elevated urinary levels of creatine and allantoin, resorption of ova in females, and occasionally sterility in the males.

Vitamin K studies reported by WOOLLEY (1945) indicated that vaginal hemorrhages and resorptions could be corrected by feeding a source of K to mice that had previously received only dL-alpha-tocopherol quinone, an analogue of K. Other symptoms include prolonged clotting time and increased hemorrhages as a result of decreased blood levels of prothrombin. Inasmuch as this vitamin is normally synthesized by microorganisms in the gut, deficiency symptoms do not usually occur unless (1) sulfonamide drugs are fed, (2) coprophagy is prevented, (3) bile acids are diverted from the intestinal tract, or (4) vitamin K antagonists are fed (LOOSLI, 1963).

WATER-SOLUBLE VITAMINS

Thiamine, or B_1, as it is commonly referred to, has received a great deal of attention related to deficiency symptoms. Specific requirements for normal reproduction and lactation are not known, but a report by MIRONE and

CERECEDO (1947) established the fact that 20 mg/kg of diet was adequate. Thiamine, being very sensitive to heat sterilization, is sufficiently overfortified in sterilizable diets. In chow formulated for breeder mice, the level of B_1 is usually elevated.

Convulsions, circular movements, brain hemorrhage, reduced food intake, poor growth, nonspecific muscle lesions, testicular degeneration, and early mortality are symptoms that have all been reported (MORRIS, 1947; JONES et al., 1945; LOOSLI, 1963). MORRIS (1944) states that the presence of thiamine is required for the metabolism of glycogen. Early clinical manifestation of thiamine-deficient diets is the failure to maintain an adequate caloric intake.

Riboflavin deficiency has been reported by LIPPINCOTT and MORRIS (1942), MORRIS and ROBERTSON (1943), and LANGSTON et al. (1933). Symptoms include dermatitis, dermatosis (atrophic and hyperkeratotic), keratitis, occasionally cataracts, myelin degeneration in the spinal cord, death of mice, and growth failure in the young, and KLIGLER et al. (1944) reported lowered resistance to *Salmonella* infection.

Deficiency symptoms of dietary B_6 (Pyridoxine) are known to include alopecia, poor growth of weanlings, posterior paralysis, necrotic degeneration of the tail, and hyperirritability (BECK et al., 1950). LYON et al. (1958) reported on the increased susceptibility of the I strain as compared with the C57 strain on a B_6-deficient diet. Greasy fur, running fits, and tonic and clonic convulsions were symptomatic in the I strain, with ultimate death, whereas only greasy fur and graying of the hair were reported in the C57 strain.

MORRIS and LIPPINCOTT (1941) described the deficiency symptoms of pantothenic acid as follows: generalized weight loss, hair loss, partial posterior paralysis (probably owing to nerve derangement), dermatosis, and, in black strains of mice, graying of the hair coat.

Mouse diets deficient in biotin cause symptoms quite similar to those of pantothenic acid deficiency. Symptoms include alopecia, growth failure in weanlings, and achromotrichia (loss of the color of the hair), as reported by NIELSEN and BLACK (1944).

Deficiencies of folic acid in the mouse diet have also been studied and reported by NIELSEN and BLACK (1944). They demonstrated the requirements of folic acid in the diet for growing mice. The effect of folic acid in the diet on lactation in mice has been analyzed by CERECEDO and MIRONE (1947).

JAFFEE (1950) reported on the dietary requirements of vitamin B_{12} and the requirements for reproduction and lactation. Deficiency signs in young mice were noted as retarded growth and renal atrophy.

Very little has been written concerning diets deficient in inositol, paraaminobenzoic acid, and choline, although WHITE and CERECEDO (1946) reported on kidney lesions in mice resulting from a choline deficiency.

As well as specific effects of certain nutritional deficiencies, extensive research has been devoted to the relationship of diet and susceptibility to infection. SMITH and DUBOS (1956) wrote on the susceptibility of mice to a coagulase-positive hemolytic staphylococcus as estimated by the extent and time of mortality and the number of colonies grown from liver and spleen cultures at various intervals during the experiment. Their results indicated that complete deprival of food for 36 to 48 hr immediately before infection with the staph organism resulted in increased susceptibility to the organism.

Although we are referring throughout this section to mouse feed, laboratories may elect to use a commercial diet prepared for other species (that is, guinea pigs, hamsters, dogs, etc.). In a report by GRIFFIN (1952), *Salmonella newport* and other *Salmonella* organisms were recovered from dog feed supplied to mouse breeding colonies at the time of unexplained outbreaks of *S. newport* infections in the colonies.

ZOONOSES AND ALLERGY

Very recently, research organizations have become concerned about the possibility of personnel contracting diseases from their research animals, important because some personnel may be exposed without basic knowledge of their risk. This factor brings up the problems of the liability of the research laboratory.

Other significant problems encountered are the serious allergic responses suffered by some sensitive individuals after repeated exposures to animal dander and fluids. In this chapter, we briefly discuss these problems and offer possible controls; however, much additional research needs to be done in these areas.

Zoonoses

In 1964, the Institute of Laboratory Animal Resources sponsored a symposium entitled "Infections of Laboratory Animals Potentially Dangerous to Man." This important symposium emphasized the potential dangers to personnel working with animals. Although laboratory mice in the controlled ecological classification (i.e., GF, DF, and SPF) are probably not carrying any zoonotic disease, the danger of infection from conventional mice being shipped in from a variety of outside sources is a real threat.

The Biological Safety Committee of the American Association for Laboratory Animal Science has published the booklet *Laboratory Infections Bibliography*. Extracted from this booklet are the following references, which discuss diseases known to be carried by mice and rats and which are potential zoonoses.

Armstrong, C., and J. W. Hornibrook. "Choriomeningitis Virus Infection Without Central Nervous System Manifestations: Report of a Case." *U.S. Public Health Rep.* **56**: 907–909, 1941.

Armstrong, C., and R. D. Lillie. "Experimental Lymphocytic Choriomeningitis of Monkeys and Mice Produced by Virus Encountered in Studies of 1933 St. Louis Encephalitis Epidemic." *U.S. Public Health Rep.* **49**: 414–432, 1936.

Borgen, L. O. "Infection with *Actinomyces muris ratti* After a Rat Bite." *Acta Pathol. Microbiol. Scand.* **25**: 161–166, 1948.

Brown, T. McP., and J. C. Nunemaker. "Rat Bite Fever. A Review of the American Cases with Reevaluation of Etiology: Report of Cases." *Bull. Johns Hopkins Hosp.* **70**: 201–327, 1942.

Clearkin, P. A. "Case of Rat Bite Fever Contracted in the Laboratory." *Kenya East African Med. J.* **5**: 196–200, 1928–29.

Hamburger, M., and H. C. Knowles. "*Streptobacillus moniliformis* Infection Complicated by Acute Bacterial Endocarditis. Report of a Case in a Physician Following Bite of Laboratory Rat." *Arch. Internal Med.* **92**: 216–220, 1953.

Hayes, G. S., and T. L. Hartman. "Lymphocytic Choriomeningitis. Report of a Laboratory Infection." *Bull. Johns Hopkins Hosp.* **73**: 275–286, 1943.

Levaditi, C., S. Nicolau, and P. Poinclaux. "Infection with *Streptobacillus moniliformis*." *Compt. Rend.* **180**: 118, 1926; *Presse Med.* **34**: 340, 1925.

Milzer, A. "Neurotropic Virus Infections in Chicago 1929–1941. Nine Cases of Lymphocytic Choriomeningitis." *Proc. Soc. Exp. Biol. Med.* **54**: 279–282, 1943.

Milzer, A., and S. O. Levinson. "Laboratory Infection with Lymphocytic Choriomeningitis. A Two Year Study of Antibody Response." *J. Am. Med. Assoc.* **120**: 27–30, 1942.

Parker, F., Jr., and N. P. Hudson. "Infection with *Streptobacillus moniliformis*." *Am. J. Pathol.* **2**: 357, 1926.

Place, E. H., and L. E. Sutton. "Infection with *Streptobacillus moniliformis*." *Arch. Internal Med.* **54**: 659, 1934.

Ribley, H. S., and H. M. Van Sant. "Rat Bite Fever Acquired from a Dog." *J. Am. Med. Assoc.* **102**: 1917–1921, 1934.

Rivers, T. M., and T. F. McN. Scott. "Meningitis in Man Caused by a Filterable Virus. II. Identification of Etiological Agent." *J. Exp. Med.* **63**: 415–432, 1936.

Smadel, J. E., R. H. Green, R. M. Paltauf, and T. A. Gonzales. "Lymphocytic Choriomeningitis: Two Human Fatalities Following an Unusual Febrile Illness." *Proc. Soc. Exp. Biol. Med.* **49**: 683–686, 1942.

Wooley, J. G., F. D. Stimpert, J. F. Kessel, and C. Armstrong. "A Study of Human Sera Antibodies Capable of Neutralizing the Virus of Lymphocytic Choriomeningitis." *U.S. Public Health Rep.* **54**: 938–944, 1939.

The booklet is filled with references pertaining to infections of laboratory workers with everything from infectious hepatitis in chimpanzees to accidental vaccinia inoculations. Anyone who doubts the importance of laboratory infections—either zoonotic, artificially zoonotic (from animals purposely infected for experimental purposes), or bench accidents—would do well to study this safety committee report carefully. The incidence of murine virus antibody in humans in contact with laboratory mice is of great interest and is currently under investigation by Dr. R. W. Tennant of our laboratory. In a brief note published in 1967, TENNANT et al. observed that the sera of both contact and noncontact groups showed antibody to Reovirus type 3 and Sendai viruses. A high incidence of antibody to Theiler's GDVII virus was also found in both groups.

HARTLEY et al. (1964) have demonstrated the presence of antibody to mouse hepatitis virus in various human population groups they studied. This finding, of course, needs much amplification and study of the socioeconomic aspects of the test groups and a determination of the epidemiology of these virus infections in humans, if they actually do occur. Potential problems involving murine viruses and laboratory personnel need to be more clearly defined.

In many zoonoses and artificial zoonoses, the external parasites with which the animals may be infected are of great importance as vectors. Infected arthropods may transfer infections from an animal to a caretaker or scientist and from animal to animal, which again emphasizes the importance of keeping animal colonies as free as possible from the broad spectrum of external parasites.

Other diseases of mice, such as *Salmonella typhimurium*, can cause epidemics of food poisoning if they somehow become involved with food preparation. It is not a good idea to use bedding from mouse pans as fertilizer on home plants, gardens, or livestock areas without first autoclaving it. This practice is common in some areas and may account for some of the mysterious epidemiology encountered with *Salmonella* food poisonings. All culture material, media, and so forth, involved in testing for such organisms should be autoclaved before discarding to protect service personnel from accidental infections.

Perhaps one of the most significant dangers to personnel working with mice is lymphocytic choriomeningitis virus (MAURER, 1964). It apparently infects all species, including man. It may, in its most acute form, cause an encephalomyelitis that can be fatal. It is highly prevalent in wild rodents and may be present in a mouse colony in a subclinical or latent form that is not detectable by the personnel working with the animals.

It may be passed as a contaminant of tissues or sera inoculum transported from one laboratory to another. The usual form of the disease in humans is normally an influenza disease, which probably accounts for the large percentage of humans who show antibody to the virus but have no history of encephalitic symptoms. Obviously, the disease is a hazard to interpretation of experimental data, as well as to personnel. Excreta may contain the virus and it may be transmitted vertically from dam to offspring; therefore, derivation procedures do not automatically eliminate it. Testing for antibody or intracerebral inoculations of control mice should be done routinely to establish colony status.

In summary, the potential probability of zoonotic infections depends a great deal on the ecological classification of the animals. In conventional colonies, where most problems will no doubt arise, general sanitation, pest and parasite control, and adequate monitoring procedures will help control the problems and may also relieve the laboratories of some liability.

Allergy to Laboratory Mice

A constant problem in people who have worked for long periods of time with laboratory mice is one of allergic sensitization to the mice. It varies greatly in intensity from a mild hayfeverlike syndrome to an acute anaphylactic response when an antigenic stimulus occurs. Individuals in our laboratory seem to vary rather remarkably on the amount and type of antigenic stimulation needed to trigger a serious episode. One individual gets along fairly well unless he suffers a bite or some other very direct contact. Others are sensitive even to clothing that has recently been in mouse housing areas. Obviously, this factor produces extreme distress to a person whose experimental work involves mice. Often this sensitivity develops when it is much too late for a complete change in research interests. When it happens at the technical support personnel level, it is probably best for such individuals to be changed to other areas of work. PATTERSON (1964) discusses this problem by dividing the types of allergic reactions into the following categories: allergic rhinitis; allergic conjunctivitis; bronchial asthma; atopic dermatitis; and, of course, anaphylactic shock. He believes the primary source of the antigen is the dandruff or the epithelial debris from the skin, that these antigens are introduced into the body by inhalation, and that the reaction occurs in the respiratory mucosa. In addition to this, fecal matter and urine that cling to dust particles may be associated with allergies to mice. Even the tissues and serum being used in centrifuges and blenders, and so forth, may develop aerosols potentially dangerous to reactions and which may sensitize others. Many mouse colonies are run in such a haphazard and unsanitary way that the antigenicity is accentuated rather than minimized.

Air control systems are important. Allergic individuals who can go into well-ventilated areas with proper air changes and work quite comfortably are completely unable to work in other areas with poor air control. In studies under way in this laboratory, particles are counted in a size range from 0.1 to 100 μ in different areas in an attempt to correlate the particle load with the type of housing and with the type of air system. The results are quite predictable. Filter-capped cage areas have much lower particle counts than cleaning areas and unfiltered animal rooms. Barrier areas have relatively low counts. The main consideration here is whether the number of allergic people being sensitized can be markedly reduced by use of filter-cap cages and better sanitation methods; we believe that it can be.

Treatments for the allergies, once developed, are expensive, time-consuming, and not very effective. However, it is probably a sound idea to call in a medical consultant when the problem arises, if it is possible.

Recommended Reading for Chapter 3

Argonne National Laboratory. *Laboratory Animal Science: A Review of the Literature for January, February, and March, 1966.* ANL-7300, Issue No. 1 (Argonne, Illinois: Laboratory Animal Information Center, Biological and Medical Research Division, June 1966).

Argonne National Laboratory. *Laboratory Animal Science: A Review of the Literature for April, May, and June, 1966.* ANL-7300, Issue No. 2 (Argonne, Illinois: Laboratory Animal Information Center, Biological and Medical Research Division, November 1966).

Conalty, M. L., ed. *Husbandry of Laboratory Animals*, 3rd Symposium of the International Committee on Laboratory Animals (New York: Academic Press, 1967).

Harris, R. J. C., ed. *The Problems of Laboratory Animal Disease* (New York: Academic Press, 1962).

Journal of the National Cancer Institute, Vol. 20, No. 5, May 1958. Papers presented at the Armed Forces Institute of Pathology, Postgraduate Course (Washington, D.C., 1954).

Kirk, Robert W., ed. *Current Veterinary Therapy: Small Animal Practice—1966-1967* (Philadelphia: W. B. Saunders Co., 1966).

Klieneberger-Nobel, E. *Pleuropneumonia-Like Organisms (PPLO). Mycoplasmataceae* (New York: Academic Press, 1962), 157 pp.

Lane-Petter, W., ed. *Animals for Research: Principles of Breeding and Management* (New York: Academic Press, 1963), 531 pp.

Riley, W. F., Jr., K. W. Smith, and R. J. Flynn, eds. *Year Book of Veterinary Medicine*, Vols. 1 (1963), 2 (1964), and 3 (1966) (Chicago: Year Book Medical Publishers, Inc.).

CHAPTER 3 REFERENCES

Adler, J. H. "Effects of Early Experience and Differential Housing on Behavior and Susceptibility of Gastric Erosions in the Rat." *J. Comp. Physiol. Psychol.* **60**: 223–238, 1965.

———. "Aspects of Stress in Animals," in *Husbandry of Laboratory Animals*, 3rd Symposium of the International Committee on Laboratory Animals, ed. by M. L. Conalty (New York: Academic Press, 1967), pp. 239–254.

Angevine, D. M., and J. A. Furth. "A Fatal Disease of Middle-Aged Mice Characterized by Myocarditis with Haemorrhage in the Pleural Cavity." *Am. J. Pathol.* **19**: 187–195, 1943.

Baker, H. J., and J. R. Lindsey. "Latent Bartonellosis: A Complicating Factor in Animal Experimentation." *Exp. Hematol.* **14**: 53–65, 1967.

Ball, C. R., and W. L. Williams. "Spontaneous and Dietary-Induced Cardiovascular Lesions in DBA Mice." *Anat. Record* **152**: 199–209, 1965.

Ball, C. R., W. L. Williams, and J. M. Collum. "Cardiovascular Lesions in Swiss Mice Fed a High-Fat Low-Protein Diet with and without Betaine Supplementation." *Anat. Record* **145**: 49–59, 1963.

Baron, L. S., and S. B. Formal. "Immunization Studies with Living Vaccine of *Salmonella typhimurium*." *Proc. Soc. Exp. Biol. Med.* **104**: 565–567, 1960.

Beard, H. H., and E. Pomerene. "Studies in the Nutrition of the White Mouse. X. The Experimental Production of Rickets in Mice." *Am. J. Phys.* **89**: 54, 1929.

Beck, E. M., P. F. Fenton, and G. R. Cowgill. "The Nutrition of the Mouse. IX. Studies on Pyridoxine and Thiouracil." *Yale J. Biol. Med.* **23**: 190, 1950.

Bell, J. M., and J. D. Erfle. "The Requirements for Potassium in the Diet of the Growing Mouse." *Can. J. Animal Sci.* **38**: 145, 1958.

Black, Paul H., Janet W. Hartley, and Wallace P. Rowe. "Isolation of a Cytomegalovirus from African Green Monkey." *Proc. Soc. Exp. Biol. Med.* **112**: 601–605, 1963.

Bleby, John. "Specific-Pathogen-Free Animals," in *The UFAW Handbook on the Care and Management of Laboratory Animals*, 3rd ed., ed. by W. Lane-Petter *et al.* (Baltimore: The Williams & Wilkins Co., 1967), p. 212.

Brennan, Patricia C., Thomas E. Fritz, and Robert J. Flynn. "*Pasteurella pneumotropica:* Cultural and Biochemical Characteristics, and Its Association with Disease in Laboratory Animals." *Lab. Animal Care* **15**: 307–311, 1965.

Brennan, Patricia C., T. E. Fritz, R. J. Flynn, and C. M. Poole. "*Citrobacter freundii* Associated with Diarrhea in Laboratory Mice." *Lab. Animal Care* **15**: 266–275, 1965.

Brick, J. O. Unpublished Data, 1966.

Briody, B. A. "Responses of Mice to Ectromelia and Vacinnia Viruses." *Bacteriol. Rev.* **23**: 61–95, 1959.

Brodsky, Isadore, and Wallace P. Rowe. "Chronic Subclinical Infection with MSGV." *Proc. Soc. Exp. Biol. Med.* **99**: 654–655, 1958.

Bryan, W. L., and K. E. Mason. "Vitamin E Deficiency in the Mouse." *Am. J. Physiol.* **131**: 263, 1940.

Burrows, William. *Textbook of Microbiology* (Philadelphia: W. B. Saunders Co., 1963), pp. 972–973, 1079–1080.

Butenko, Z. A. "On the Possibility of Transmitting Leukosis by Bedbugs." *Probl. Virol. (English Transl.)* **3**: 231–235, 1958.

Carnochan, F. G., and C. N. W. Cumming. "Immunization against Salmonella Infection in a Breeding Colony of Mice." *J. Infect. Diseases* **90**: 242, 1952.

Cerecedo, L. R., and L. Mirone. "The Beneficial Effect of Folic Acid (Lactobacillus casei factor) on Lactation in Mice Maintained on Highly Purified Diets." *Arch. Biochem.* **12**: 154, 1947.

Cerecedo, L. R., and L. J. Vinson. "Growth, Reproduction and Lactation in Mice on Highly Purified Diets, and the Effect of Folic Acid Concentrates on Lactation." *Arch. Biochem.* **5**: 157, 1944.

Congdon, C. C., D. A. Gardiner, and M. A. Kastenbaum. "Reduced Secondary Disease Mortality in Mouse Radiation Chimeras." *J. Natl. Cancer Inst.* **38**: 541–548, 1967.

Congdon, C. C., M. A. Kastenbaum, and D. A. Gardiner. "Factors Affecting Mortality from Secondary Disease in Mouse Radiation Chimeras." *J. Natl. Cancer Inst.* **35**: 227–234, 1965.

Congdon, C. C., and I. S. Urso. "Homologous Bone Marrow in the Treatment of Radiation Injury in Mice." *Am. J. Pathol.* **33**: 749–767, 1957.

Cook, Iran. "Reovirus Type 3 Infection in Laboratory Mice." *Australian J. Exp. Biol. Med. Sci.* **41**: 651, 1963.

Crystal, M. M. (a) "The Infective Index of the Spine Rat Louse, *Polyplax spinulosa* (Burmeister), in the Transmission of *Hemobartonella muris* (Mayer) of Rats." *J. Econ. Entomol.* **52**: 534–544, 1959. (b) "Extrinsic Incubation Period of *Hemobartonella muris* in the spined rat louse, *Polyplax spinulosa*." *J. Bacteriol.* **77**: 511, 1959.

Day, H. G., and B. E. Skidmore. "Some Effects of Dietary Zinc Deficiency in the Mouse." *J. Nutr.* **33**: 27, 1947.

Deringer, M. K. "Necrotizing Arteritis in Strain BL/DE Mice." *Lab. Invest.* **8**: 1461–1465, 1959.

Deringer, M. K., T. B. Dunn, and W. E. Heston. "Results of Exposure to Strain C3H Mice to Chloroform." *Proc. Soc. Exp. Biol. Med.* **83**: 474–479, 1953.

DeVries, M. J., and O. Vos. "Delayed Mortality of Radiation Chimeras. A Pathological and Hematological Study." *J. Natl. Cancer Inst.* **23**: 1403–1439, 1959.

Erway, L., L. S. Hurley, and A. Fraser. "Neurological Defect: Manganese in Phenocopy and Prevention of a Genetic Abnormality of Inner Ear." *Science* **152**: 1766–1777, 1966.

Ewing, W. H. "Enterobacteriaceae; Biochemical Methods for Group Differentiation." *U.S. Public Health Serv. Publ.* **734**: 1–32, 1962.

Fisher, E. R., and L. Kilham. "Pathology of a Pneumotropic Virus Recovered from C3H Mice Carrying the Bittner Milk Agent." *Arch. Pathol.* **55**: 14–19, 1953.

Flynn, R. J. "Studies on the Etiology of Ringtail of Rats." *Proc. Animal Care Panel* **9**: 155–160, 1959.

———. "The Diagnosis and Control of Ectromelia Infection of Mice." *Lab. Animal Care* **13**: 130–136, 1963.

Freundt, E. A. "Arthritis Caused by *Streptobacillus moniliformis* and Pleuropneumonia-like Organisms in Small Rodents." *Lab. Invest.* **8**: 1358–1375, 1959.

Fry, R. J. M., K. Hamilton, and H. Lisco. "An Unusual Spontaneous Cardiac Lesion of Unknown Etiology in Mice." *Federation Proc.* **19**: 109, 1960.

Furth, J., and J. Moshman. "On the Specificity of Hypervolemia and Congestive Changes in the Tumor-Bearing Mice." *Cancer Res.* **11**: 543–551, 1951.

Goettsch, M. A. "Alpha-tocopherol Requirement of the Mouse." *J. Nutr.* **23**: 513, 1942.

Green, Earl L., ed. *Biology of the Laboratory Mouse*, 2nd ed. (New York: McGraw-Hill Book Company, 1966), 706 pp.

Griffin, C. A. "A Study of Prepared Feeds in Relation of Salmonella Infection in Animals." *J. Am. Vet. Med. Assoc.* **121**: 197, 1952.

Gruneberg, H. *The Genetics of the Mouse*, 2nd ed. (The Hague: Martinus Nijhoff, 1952), 650 pp.

Hall, C. E., E. Cross, and O. Hall. "Amyloidosis and Other Pathologic Changes in Mice Exposed to Chronic Stress." *Texas Rept. Biol. Med.* **18**: 205–213, 1960.

Hartley, J. W., and W. P. Rowe. "A New Mouse Virus Apparently Related to the Adenovirus Group." *Virology* **11**: 645–647, 1960.

Hartley, Janet W., Wallace P. Rowe, H. H. Bloom, and Horace C. Turner. "Antibodies to Mouse Hepatitis Viruses in Human Serums." *Proc. Soc. Exp. Biol. Med.* **115**: 414–418, 1964.

Hartley, Janet W., Wallace P. Rowe, and Robert J. Huebner. "Serial Propagation of the Guinea Pig SGV in Tissue Culture." *Proc. Soc. Exp. Biol. Med.* **96**: 281–285, 1957.

Heston, W. E. "The Genetic Aspects of Lung Tumors in Mice," in *Lung Tumours of Animals*, ed. by L. Severi (Perugia: Division of Cancer Research, 1966), pp. XLIII–LVI.

Hinton, David E., and W. Lane-Williams. "Hepatic Fibrosis Associated with Aging in Four Stocks of Mice." *J. Gerontol.* **23**: 205–211, 1968.

Hoag, W. G., H. Meier, M. Dickie, E. P. Les, and J. Dorey. "The Effect of Nutrition on Fertility of Inbred DBA/2J Mice." *Lab. Animal Care* **16**: 228–236, 1966.

Hollander, W. F. "Colonic Intussusception in the Mouse." *Am. J. Vet. Res.* **20**: 750–752, 1959.

Holt, D. "Presence of K Virus in Wild Mice in Australia." *Australian J. Exp. Biol. Med. Sci.* **37**: 183, 1959.

Horsfall, F. L., and R. G. Hahn. "Latent Virus in Normal Mice Capable of Producing Pneumonia in Its Natural Host." *J. Exp. Med.* **71**: 391, 1940.

Iida, T., and F. B. Bang. "Infection of the Upper Respiratory Tract of Mice with Influenza A Virus." *Am. J. Hyg.* **77**: 169, 1963.

Ito, S., and T. Tanaka. "Bacterial Investigations in Commercial Mice." *Bull. Exp. Animals* **1**: 27–30, 1952.

Jaffee, W. F. "Reproduction of Mice Kept on Rations Low in Vitamin B_{12}." *Arch. Biochem.* **27**: 464, 1950.

Jones, J. H., C. Foster, F. Dorfman, and G. L. Hunter. "Effects on the Albino Mouse of Feeding Diets Very Deficient in Each of Several Vitamin B Factors (Thiamine, Riboflavin, Pyridoxine, Pantothenic Acid)." *J. Nutr.* **29**: 127, 1945.

Kilham, L., and H. W. Murphy. "A Pneumotropic Virus Isolated from C3H Mice Carrying the Bittner Milk Agent." *Proc. Soc. Exp. Biol. Med.* **82**: 133, 1953.

King, E. O., M. K. Ward, and D. F. Raney. "Two Simple Media for the Demonstration of Pyocyanin and Fluorescin." *J. Lab. Clin. Med.* **44**: 301–307, 1954.

Kligler, I. J., K. Guggenheim, and E. Buechler. "Relation of Riboflavin Deficiency to Spontaneous Epidemics of *Salmonella* in Mice." *Proc. Soc. Exp. Biol. Med.* **57**: 132, 1944.

Kraft, L. M. "Observations on the Control and Natural History of Epidemic Diarrhea of Infant Mice." *Yale J. Biol. Med.* **31**: 121–137, 1958.

——. "Two Viruses Causing Diarrhea in Infant Mice," in *The Problems of Laboratory Animal Disease*, ed. by R. J. C. Harris (New York: Academic Press, 1962), pp. 115–130.

Krakower, Cecil, and Luis M. Gonzalez. "Mouse Leprosy." *Science* **86**: 617, 1937.

Kuhl, I. "Uber Allergisch-Hyperergische Erscheinunger bei Mausen nach β-napthylaminebehandlung Gekennzeichnet durch *Periarteritis nodosa*." *Virchows Arch. Pathol. Anat. Physiol. Klin. Med.* **328**: 49–67, 1956.

Lainson, R., P. C. C. Garnham, R. Killick-Kendrick, and R. G. Bird. "Nosematosis, a Microsporidial Infection of Rodents and Other Animals, Including Man." *Brit. Med. J.* **2**: 470–472, 1964.

Lane-Petter, W., ed. *Animals for Research* (New York: Academic Press, 1963), pp. 54–55, 61–66.

Langston, W. C., P. L. Day, and K. W. Cosgrove. "Cataract in the Albino Mouse Resulting from a Deficiency of Vitamin G (B_2)." *Arch. Ophthalmol.* **10**: 508, 1933.

Lippincott, S. W., and H. P. Morris. "Pathologic Changes Associated with Riboflavin Deficiency in the Mouse." *J. Natl. Cancer Inst.* **2**: 601, 1942.

Loosli, J. K. "Primary Signs of Nutritional Deficiencies of Laboratory Animals." *J. Am. Vet. Med. Assoc.* **142**: 1001, 1963.

Lowenthal, K. "Nekrotisierende aortitis mit aortenruptur bei einer Maus." *Virchows Arch. Pathol. Anat. Physiol. Klin. Med.* **265**: 424, 1927.

Lyon, J. B., H. L. Williams, and E. A. Arnold. "The Pyridoxine-Deficient State in Two Strains of Inbred Mice." *J. Nutr.* **66**: 261, 1958.

Maurer, Fred D. "Lymphocytic Choriomeningitis." *Lab. Animal Care* **14**: 415–419, 1964.

McCarthy, P. T., and L. R. Cerecedo. "Vitamin A Deficiency in the Mouse." *J. Nutr.* **46**: 361, 1952.

Medearis, Donald N., Jr. (a) "Mouse Cytomegalovirus Infection. II. Observation During Prolonged Infections." *Am. J. Hyg.* **80**: 103–112, 1964. (b) "Mouse Cytomegalovirus Infection. III. Attempts to Produce Intrauterine Infections." *Am. J. Hyg.* **80**: 113–120, 1964.

Meier, H., R. C. Allen, and W. C. Hoag. "Spontaneous Haemorrhagic Diathesis in Inbred Mice Due to Single and Multiple 'Prothrombin Complex' Differences." *Blood* **19**: 501–514, 1962.

Mirone, L., and L. R. Cerecedo. "The Beneficial Effect of Xanthopterin on Lactation, and of Biotin on Reproduction and Lactation, in Mice Maintained on Highly Purified Diets." *Arch. Biochem.* **15**: 324, 1947.

Miller, C. P., C. W. Hammond, M. Tompkins, and G. Shorter. "The Treatment of Postirradiation Infection with Antibiotics; and Experimental Study on Mice." *J. Lab. Clin. Med.* **39**: 462–479, 1952.

Milzer, A. "Studies on the Transmission of Lymphocytic Choriomeningitis Virus by Arthropods." *J. Infect. Diseases* **70**: 152, 1942.

Milzer, A., and S. O. Levinson. "Active Immunization of Mice with Ultraviolet-Inactivated Lymphocytic Choriomeningitis Virus Vaccine and Results of Immune Serum Therapy." *J. Infect. Diseases* **85**: 251–255, 1949.

Morris, H. P. "Review of the Nutritive Requirements of Normal Mice for Growth, Maintenance, Reproduction and Lactation." *J. Natl. Cancer Inst.* **5**: 115–141, 1944.

———. "Vitamin Requirements of the Mouse." *Vitamins Hormones* **5**: 175, 1947.

Morris, H. P., and S. W. Lippincott. "The Effect of Pantothenic Acid on Growth and Maintenance of Life in Mice of the C$_3$H Strain." *J. Natl. Cancer Inst.* **2**: 29, 1941.

Morris, H. P., and W. V. B. Robertson. "Growth Rate and Number of Spontaneous Mammary Carcinomas and Riboflavin Concentration of Liver, Muscle, and Tumor of C$_3$H Mice as Influenced by Dietary Riboflavin." *J. Natl. Cancer Inst.* **3**: 479, 1943.

Nelson, J. B. (a) "Infectious Catarrh of Mice. I. A Natural Outbreak of the Disease." *J. Exp. Med.* **65**: 833–842, 1937. (b) "Infectious Catarrh of Mice. II. The Detection and Isolation of Coccobacilliform Bodies." *J. Exp. Med.* **65**: 843–850, 1937. (c) "The Etiological Significance of the Coccobacillary Bodies." *J. Exp. Med.* **65**: 851–860, 1937.

———. "The Transmission of a Communicable Disease of Mice from Naturally Infected Rats and the Nature of the Causal Agent in Experimentally Infected Mice." *J. Exp. Med.* **84**: 7–14, 1946.

Nelson, J. B., and G. R. Collins. "The Establishment and Maintenance of a Specific Pathogen Free Colony of Swiss Mice." *Proc. Animal Care Panel* **11**: 65–70, 1961.

Nielsen, E., and A. Black. "Biotin and Folic Acid Deficiencies in the Mouse." *J. Nutr.* **28**: 203, 1944.

Niven, Janet S. F. "Hepatotropism in Virus Infections of Mice," in *Pathology of Laboratory Rats and Mice*, ed. by E. Cotchin and F. J. C. Roe (Philadelphia: F. A. Davis Company, 1967), Chap. 18, pp. 586–587.

Noda, E. "Tetraethyl Lead Poisoning." *Fukuoka Igaku Zassi* **49**: 2779–2789, 1958.

Pappenheimer, A. M. "Muscular Dystrophy in Mice on Vitamin E-Deficient Diet." *Am. J. Pathol.* **18**: 169, 1942.

———. "Myocarditis and Pulmonary Arteritis Associated with the Presence of Rickettsia-Like Bodies in Polymorphonuclear Leukocytes." *J. Natl. Cancer Inst.* **20**: 921–931, 1958.

Parker, John C., Raymond W. Tennant, and Thomas G. Ward. "Enzootic Sendai Virus Infections in Mouse Breeder Colonies Within the United States." *Science* **146**: 936–938, 1964.

———. "Prevalence of Viruses in Mouse Colonies." *Natl. Cancer Inst. Monogr.* **20**: 25–36, 1966.

Patterson, Roy. "The Problem of Allergy to Laboratory Animals." *Lab. Animal Care* **14**: 466–469, 1964.

Perman, Victor, and Martin E. Bergeland. "A Tularemia Enzootic in a Closed Hamster Breeding Colony." *Lab. Animal Care* **17**: 563–568, 1967.

Rees, R. J. W. "Enhanced Susceptibility of Thymectomized and Irradiated Mice to Infection with *Mycobacterium leprae*." *Nature* **211**: 657–658, 1966.

Rees, R. J. W., M. F. R. Waters, A. G. M. Weddell, and E. Palmer. "Experimental Lepromatous Leprosy." *Nature* **215**: 599–602, 1967.

Richter, C. B., and J. O. Brick. "Observations on a Leukemia-like Disease in C57BL/6 Mice." *Exp. Hematol.* **14**: 74–76, 1967.

Riley, V. "*Eperythrozoon coccoides*." *Science* **146**: 921–923, 1964.

Riley, V., F. Lilly, E. Huerto, and D. Bardell. "Transmissible Agent Associated with 26 Types of Experimental Mouse Neoplasms." *Science* **132**: 545–547, 1960.

Rowe, W. P., J. W. Hartley, and R. J. Huebner. "Polyoma and Other Indigenous Mouse Viruses," in *The Problems of Laboratory Animal Disease*, ed. by R. J. C. Harris (New York: Academic Press, 1962), pp. 131–142.

Rupple, B. M. "Congenital Polycystic Disease of the Kidney Occurring in Mice." *J. Natl. Cancer Inst.* **15**: 1183–1194, 1955.

Salaman, M. H. "Virus-Induced Lymphoma in Mice," in *Pathology of Laboratory Rats and Mice*, ed. by E. Cotchin and F. J. C. Roe (Philadelphia: F. A. Davis Company, 1967), Chap. 19, pp. 613–637.

Schneider, H. A. "Nutritional and Genetic Factors in the Natural Resistance of Mice to Salmonella Infections." *Ann. N.Y. Acad. Sci.* **66**: 337–347, 1956.

———. "Nutritional Factors in Host Resistance." *Bacteriol. Rev.* **24**: 186–191, 1960.

Shechmeister, I. L. "Multiplication of *Salmonella enteritidis* in X-irradiated and Normal Mice." *Lab. Animal Care* **17**: 585–588, 1967.

Shechmeister, I. L., L. J. Paulissen, and M. Fishman. "Sublethal Total Body X-Radiation and Susceptibility of Mice to *Salmonella enteritidis* and *Escherichia coli*." *Proc. Soc. Exp. Biol. Med.* **83**: 205–209, 1953.

Shepard, Charles C., and Charles C. Congdon. "Increased Growth of *Mycobacterium leprae* in the Foot Pads of Thymectomized-Irradiated Mice." *Exp. Hematol.* **15**: 119, 1968.

Shils, M. E., and E. V. McCollum. "Further Studies on the Symptoms of Manganese Deficiency in the Rat and Mouse." *J. Nutr.* **26**: 1, 1943.

Sidman, Richard L., Margaret C. Green, and Stanley H. Appel. *Catalog of the Neurological Mutants of the Mouse* (Cambridge: Harvard University Press, 1965), p. 82.

Silberberg, M., and R. Silberberg. "Osteoarthrosis in Mice Fed Diets Enriched with Animal or Vegetable Fat." *Arch. Pathol.* **70**: 385, 1960.

Simmons, M. L., and John Franklin. "Acute Radiation Mortality in Mice Inoculated with Pigmented or Non-Pigmented Pseudomonas." *Res. Vet. Sci.* **8**: 71–73, 1967.

Simmons, M. L., and L. H. Smith. "An Anesthetic Unit for Small Laboratory Animals." *J. Appl. Physiol.* **25**: 324–325, 1968.

Simmons, M. L., H. E. Williams, and Everline B. Wright. "Therapeutic Value of the Organic Phosphate Trichlorfon Against *Syphacia obvelata* in Inbred Mice." *Lab. Animal Care* **15**: 382–385, 1965.

Simmons, M. L., C. B. Richter, J. A. Franklin, and R. W. Tennant. "Prevention of Infectious Diseases in Experimental Mice." *Proc. Soc. Exp. Biol. Med.* **126**: 830–837, 1967.

Simon, H. J. *Attenuated Infection* (Philadelphia: J. B. Lippincott Co., 1960), p. 349.

Smith, J. M., and R. J. Dubos. "The Effect of Nutritional Disturbances on the Susceptibility of Mice to Staphylococcal Infections." *J. Exp. Med.* **103**: 119, 1956.

Smith, Kendall O., and Lucy Rasmussen. "Morphology of Cytomegalovirus." *J. Bacteriol.* **85**: 1319–1325, 1963.

Smith, Margaret G. "The Salivary Gland Viruses of Man and Animals," in *Progress in Medical Virology*, Vol. 2, ed. by E. Berger and J. L. Melnick (New York: Hafner Publishing Co., Inc., 1959), pp. 171–202.

Stanley, N. F., D. C. Dorman, and Joan Ponsford. "Studies on the Pathogenesis of a Hitherto Undescribed Virus Producing Unusual Symptoms in Suckling Mice." *Australian J. Exp. Biol. Med. Sci.* **31**: 147, 1953.

Stanley, N. F., P. J. Leak, M. N.-I. Walters, and R. A. Joske. "Murine Infection with Reovirus. II. The Chronic Disease Following Reovirus Type 3 Infection." *Brit. J. Exp. Pathol.* **45**: 142, 149, 1963.

Stewart, H. L., Katharine C. Snell, and William V. Hare. "Histopathogenesis of Carcinoma Induced in the Glandular Stomach of C57BL Mice by the Intramural Injection of 20-Methyl-Cholanthrene." *J. Natl. Cancer Inst.* **21**: 999–1035, 1958.

Stoenner, H. G., E. F. Grimes, F. B. Thrailkill, and E. Davis. "Elimination of *Leptospira ballum* from a Colony of Swiss Albino Mice by Use of Chlortetracycline Hydrochloride." *Am. J. Trop. Med. Hyg.* **7**: 423–426, 1958.

Storer, John B. "Longevity and Gross Pathology at Death in 22 Inbred Mouse Strains." *J. Gerontol.* **21**: 404–409, 1966.

Syverton, J. T., and R. G. Fischer. "The Cockroach as an Experimental Vector of the Virus of Spontaneous Mouse Encephalomyelitis (Theiler)." *Proc. Soc. Exp. Biol. Med.* **74**: 296–298, 1950.

Tennant, Raymond W. "Taxonomy of Murine Viruses." *Natl. Cancer Inst. Monogr.* **20**: 47–53, 1966.

Tennant, R. W., J. C. Parker, and T. G. Ward. "Studies on the Natural History of Pneumonia Virus of Mice." *Bacteriol. Proc.* 125, 1964.

———. "Respiratory Virus Infections of Mice." *Natl. Cancer Inst. Monogr.* **20**: 93–104, 1966.

Tennant, R. W., R. K. Reynolds, and K. R. Layman. "Incidence of Murine Virus Antibody in Humans in Contact with Experimental Animals." *Exp. Hematol.* **14**: 76, 1967.

Thung, P. J. "Senile Amyloidosis in Mice." *Gerontologia* **1**: 259–279, 1957.

Trentin, J. J. "Tolerance and Homologous Disease in Irradiated Mice Protected with Homologous Bone Marrow." *Ann. N.Y. Acad. Sci.* **73**: 799–810, 1958.

Tyzzer, E. E. "A Fatal Disease of the Japanese Waltzing Mouse Caused by a Spore-Bearing Bacillus (*Bacillus piliformis*, N. sp.)." *J. Med. Res.* **37**: 307–338, 1917.

Upton, A. C., J. W. Conklin, G. E. Cosgrove, W. D. Gude, and E. B. Darden. "Necrotizing Polyarteritis in Aging RF Mice." *Lab. Invest.* **16**: 483–487, 1967.

Van Der Waaij, D., and C. A. Sturm. "Antibiotic Decontamination of the Digestive Tract of Mice. Technical Procedures." *Lab. Animal Care* **18**: 1–10, 1968.

Walters, M. N.-I., R. A. Joske, P. J. Leak, and N. F. Stanley. "Murine Infection with Reovirus: I. Pathology of the Acute Phase." *Brit. J. Exp. Pathol.* **44**: 427–436, 1963.

Ward, R. J. "Animals for Nutritional Research." *Carworths Collected Papers* **1**: 17–24, 1967.

Ward, Thomas G. "Natural History of Sendai Virus Infection in Mice," in *Annual Progress Report July 1, 1965–July 1, 1966* (Bethesda: Microbiological Associates, Inc., 1966), pp. 91–103.

Wensinck, F., D. W. Van Bekkum, and H. Renaud. "The Prevention of *Pseudomonas aeruginosa* Infections in Irradiated Mice and Rats." *Radiation Res.* **7**: 491–499, 1957.

White, E. A., and L. R. Cerecedo. *Proc. Am. Chem. Soc.* 23–24B, 1946 [as cited by H. P. Morris (11), 1947].

Wilgram, G. F., and D. J. Ingle. "Renal-Cardiovacsular Pathologic Changes in Aging Female Breeder Rats." *Arch. Pathol.* **68**: 690–703, 1959.

Wilson, B. J., and C. H. Wilson. "Toxin from *Aspergillus flavus:* Production on Food Material of a Substance Causing Tremors in Mice." *Science* **144**: 177–178, 1964.

Wolfe, J. M., and H. P. Slater, Jr. "Vitamin A Deficiency in the Albino Mouse." *J. Nutr.* **4**: 185, 1931.

Woolley, D. W. "Some Biological Effects Produced by Alpha-tocopherol Quinone." *J. Biol. Chem.* **169**: 59, 1945.

Yunker, Conrad E. "Infections of Laboratory Animals Potentially Dangerous to Man: Ectoparasites and Other Arthropods, with Emphasis on Mites." *Lab. Animal Care* **14**: 455–465, 1964.

Techniques

FOUR

INTRODUCTION

There are numerous excellent sources of information on a variety of techniques involving the mouse. Techniques may be defined as any procedure done directly to or with the mouse itself. Examples would be surgery, fluid administration and removal, anesthesia, radiography, drug dosage, and restraint. Special techniques include irradiation of animals with shielded spleens, kidney and splenic transplants, and others. In this chapter, we discuss the most important aspects of several broad areas within this category and give additional sources of more detailed information. We also describe techniques that may be unique or that have not been reported elsewhere.

SURGICAL PROCEDURES ON THE MOUSE

The mouse is very resistant to postsurgical infections and surgical complications. Most surgical procedures undertaken with mice involve large numbers and, therefore, must be approached in such a way that the logistics of the problem do not cause overwhelming difficulties in establishing appropriate numbers and controls. It may be a sound idea to set up mock surgical situations for control groups. The method of preparation of the mice for surgery varies with the experiment or reason for the surgery. Of course, hysterectomies, which are done to derive new germfree or pathogen-free animals, must involve meticulous aseptic surgical procedures. On the other hand, splenectomies, tumor transplants, nephrectomies, and so on, can be done with "clean" techniques. In any surgical situation, it is always desirable to achieve the maximum amount of asepsis possible under the prevailing circumstances.

Preparation

Little if any preoperative care is necessary in the mouse, unless the surgery involves the alimentary canal, and it is desirable to have it empty or specifically contaminated, unless the nutrition or toxicity is, in some specific way, very important to the experiment. In practice, most mice are quickly shaved over the incision area with either small electric clippers or an ordinary razor. The site is then quickly sprayed or flooded with Merthiolate or other germicide and the surgery proceeds.

Surgery

Special surgical techniques that involve exteriorization and X irradiation of specific organs with whole-body shielding, or shielding of specific organs with whole-body irradiation, are done routinely. When the organs must be exteriorized for some time, it is not practical to attempt complete sterility. However, with a few precautions, such as using sterile physiological saline for wetting the exteriorized organ, and keeping all reusable equipment scrupulously clean, the procedures will usually be successful.

Many people doing surgical procedures on mice tend to use surgical approaches developed for larger animals. A dorsal approach in a nephrectomy is a common procedure in the mouse. However, because of group housing and the habit of some mice to fight and wound each other, it is usually better to confine as many surgical procedures as possible to the midventral approach. This system allows the individual animals to protect their wounds while healing and it is esthetically more desirable if visitors are a problem.

Standard materials and methods should be used in the ligation of internal organs. Number 00 catgut is usually adequate for internal ligations and closing the peritoneum and other cavities. It comes with small swedged needle (Appendix) and is quite satisfactory. Michelle clip devices (Appendix) are very satisfactory, provided they are placed properly.

Postsurgical Care

Postoperative mice can be put into cages with food and water immediately after completion of the surgery. The most important aspect of mouse surgery postoperative care is to make certain the mice are placed into cages with clean bedding and fresh food and water so their immediate environment will be conducive to quick healing and a minimum of infections. The mice should be observed regularly and carefully, particularly if clips, which can become very uncomfortable to the animal, are used. Clips causing apparent discomfort can be removed and new ones applied. In some cases, prophylactic therapy may be indicated owing to latent infections and other experimental reasons. Oxytetracycline or other broad spectrum antibiotics can be placed in the drinking water or even flushed into the incision site. We do not think that this procedure is ordinarily necessary or perhaps even desirable. A recent publication on murine splenectomy reporting use of an electrocautery device instead of ligation is of interest (ERSEK, 1968). Other surgical techniques, such as cryosurgery, may be used and are probably the methods of the future. The cryosurgical approach requires only minimal ligations.

FLUID ADMINISTRATION

Intraperitoneal

Intraperitoneal administration is used for many different purposes in the mouse. Anesthetics, tumor transplants, experimental drugs, and many others are routinely injected by the intraperitoneal route. The technique is simple once the basic method of picking up and holding the mouse is mastered. We recommend gently but firmly grasping the mouse by the skin at the back of the neck and carefully gathering the slack skin between the thumb and forefinger. Caution must be taken to avoid pulling the skin too tightly and strangling the mouse. With the skin held properly, the mouse can be lifted and the tail anchored between the small finger and palm of the hand. The mouse is now in a position to be easily injected by a number of routes. To inject intraperitoneally, a $\frac{1}{2}$-inch 22- to 26-gauge needle is adequate. The size of the needle depends more on the viscosity of the material being injected than on any other consideration. The needle should be introduced on a plane forming approximately a 10° angle with the abdominal surface and slightly to the right or left of the midline. The angle is important because it is easy to insert the needle into the urinary bladder or into the intestine if the angle is too great. With some practice, this technique is easily mastered.

Intramuscular

Intramuscular injections in mice are fairly difficult because of the lack of a large muscle mass. The usual site for intramuscular injections is in the posterior thigh muscle group. A $\frac{1}{4}$-inch 24-gauge needle is adequate. The same general restraining technique can be used if only one person is doing the work. Other methods of restraint for intramuscular injections include plastic tubes with holes cut in them large enough to allow the leg and thigh to be gently pulled through and injected. Caution must be used in intramuscular injections that the needle is not pushed to deep or too hard, for it may pass completely through the muscle mass. The needle should be directed perpendicular to the sagittal plane, or pointed in a very slightly posterior direction.

Subcutaneous

Subcutaneous injections in mice are probably the easiest of all to give. The mouse can be injected subcutaneously on either the dorsal or the ventral side, and a $\frac{1}{4}$-inch 22-gauge needle is adequate for most preparations. If the quantity of injected material is fairly large and does not interfere with experimental protocol, multiple injection sites are advisable. The rate of absorption is probably reduced considerably from the intraperitoneal or intramuscular injections. As with other methods, the restraint of the animal is of great importance.

Intrathoracic

Intrathoracic injections can be made in mice with a slightly bent or curved $\frac{1}{4}$-inch 22-gauge needle. It should be inserted between the ribs at approximately the midpoint of the rib cage. Caution must be taken to insert it at an angle, thus preventing injection directly into lung tissue. Intrathoracic methods are not used routinely, and, unless there is a specific experimental reason to use the method, the intraperitoneal route is easier and the speed of absorption is similar.

Intravenous

The usual site of intravenous injection in mice is the lateral tail vein. The mice can be restrained in a number of different ways. The simplest is to pass the tail through a $\frac{1}{4}$-inch slot in a small plastic shield. If the animals have been warmed under light bulbs for approximately 10 min, the veins will have expanded and intravenous injections can be accomplished using a $\frac{1}{2}$-inch 24-gauge needle. It is a technique that requires practice, but can be done routinely once the skill is developed. As with all material injected intravenously, it should not contain extraneous material that may act as an embolism and kill the animal.

Intra-arterial

There may be times when intra-arterial injections are desirable for experimental purposes or for angiography. The mouse should be anesthetized and restrained in a dorsal recumbancy. An incision is made in the skin over the femoral artery so it can be visualized. Then injections can be made in the usual manner, using a ½-inch 24-gauge needle. Microcatheters may also be used.

Oral Administration

Polyethylene stomach tubes of approximately 20 gauge and 1.5 to 2 inches long can be forced on an 18- to 20-gauge blunted hypodermic needle attached to a syringe. With practice, the mouse can be induced to accept this in the esophagus, and fluids can be deposited directly into the stomach. Care must be taken not to introduce it into the trachea. Another method involves using a long 20-gauge hypodermic needle that has had a metal bulb soldered onto the end and a hole made in the side of the needle. This method is probably easiest and does not result in chewing of the polyethylene by the mice. There are measured dose automatic syringes available that can be used for consecutive dosing. It is helpful when using any of these methods to align the esophagus and stomach orfices by tilting the animal's head dorsally.

METHODS OF BLOOD WITHDRAWAL

Orbital Sinus

Blood can be withdrawn very conveniently from the orbital sinus of the medial canthus of the eye in unanesthetized mice. A microhematocrit tube or a capillary pipette works equally well and, by using the capillary pipette, the mouse can easily be bled out completely. The technique is not difficult, although it is probably a good idea to learn on anesthetized mice because of ease of handling and minimizing the trauma to the mice during the technique-learning phase. Persons who are adept at the procedure can draw blood samples quickly and with minimal stress to the animal (RILEY, 1960).

Decapitation

Mice can be bled by decapitation with a razor blade or scissors. However, it is esthetically undesirable; also, the blood tends to be much more contaminated with hair, bacteria, and other foreign material. We do not recommend this method of collecting blood.

Cardiac Puncture

The cardiac puncture as a method of withdrawing blood samples from mice is difficult, time-consuming (it is necessary to anesthetize the mice), and ordinarily not as productive as other methods; however, many investigators do prefer it. The technique consists of restraining the anesthetized animal on its back, disinfecting the midventral thorax, and making the thoracic penetration just anterior to the xyphoid cartilage with a 2-ml syringe with a 1-inch 25-gauge needle. Slight negative pressure is drawn with the syringe and, when proper penetration of the heart is achieved, blood will flow into the syringe. This technique requires considerable practice before it is routinely productive.

There are many other methods of blood withdrawal from mice: tail vein puncture, jugular vein puncture, tail clippings, toe clippings, and cut down incision techniques. However, we feel that the orbital method is by far the most efficient and applies much less stress on the animals.

Lymph Collection

A technique for exposing the thoracic duct in the abdominal cavity and withdrawing lymph was described by SHREWSBURY (1958). The technique involves the dorsal approach similar to that used in a left dorsal adrenalectomy. A dissecting microscope is used, and bloodless dissection exposes the thoracic duct contiguous to the posterior side of the aorta near the diaphragm. Shrewsbury also reported average lymph flow rate and cell counts per hour. The technique is not simple but can be done by determined investigators.

ASCITES FLUID COLLECTION

There are several reasons for withdrawing ascites fluids experimentally: (1) to recover ascites tumor cells; (2) to relieve ascites fluid pressure; and (3) to recover antibodies from the ascites fluids.

The technique is simply a reverse of the intraperitoneal inoculation described previously. A negative pressure is drawn on the syringe and the fluid drawn off. A polyethylene tube attached to a needle may be used and slight abdominal pressure applied to drain the fluid. It is a problem to produce large quantities of mouse serum for antibody production. A technique has been reported by MUNOZ (1957) for injecting intraperitoneally antigens mixed with Freund's adjuvant. This technique causes the development of large amounts of peritoneal fluid, which may contain specific antibody in high concentration. It seems to be an important use of ascites fluids, and, of course, withdrawal techniques are important in this procedure.

The technique may be expanded by using ascites tumor cells to increase the fluid production. If this procedure is contemplated, the tumor line should be carefully tested to make certain it is not contaminated by any of the latent murine viruses. It would be highly undesirable to include additional antigens in an antibody production experiment.

URINE COLLECTION

Rodent metabolism cages may be used to collect urine (Appendix) and feces of mice. However, if small quantities are needed, it is often possible to catch a few drops of urine simply by abruptly picking up a mouse in your left hand while deftly holding a large-mouthed test tube in your right hand. Most mice will urinate profusely immediately after being picked up, and it is possible to collect this spontaneous urine.

SPECIAL TECHNIQUES

Some special techniques of interest to investigators using mice are spleen-shielding whole-body irradiation, whole-body-shielding spleen irradiation, and various surgical grafting techniques. In this section, we describe a few of the more interesting of these techniques. In some figures, we have used rats for demonstration purposes because the larger size is easier to photograph; however, all the techniques described are applicable to mice.

A technique for exteriorizing the spleen, and shielding it from the irradiation source while giving the rest of the animal whole-body irradiation, was described by JACOBSON *et al.* in 1949. They used mice and X irradiation in their study. The spleens were exteriorized through an upper-left abdominal quadrant incision and placed in a lead box with only the pedicle receiving any appreciable irradiation. The radiation took approximately 12 min, after which the spleens were returned to the abdominal cavity and the incision closed. In Figures 4.1, 4.2, and 4.3, a similar method, used by Dr. Lawton Smith of this laboratory, is shown.

The important aspects of such surgical techniques are to exteriorize the spleen, including taking care to prevent it from drying by keeping it moist with physiological saline solutions. The time the organ is exteriorized, of course, must be minimized. Caution must be taken that the pedicle of arteries and veins supplying the organ be handled as carefully as possible to avoid any trauma that may cause splenic infarcts, and therefore interfere with the experiment.

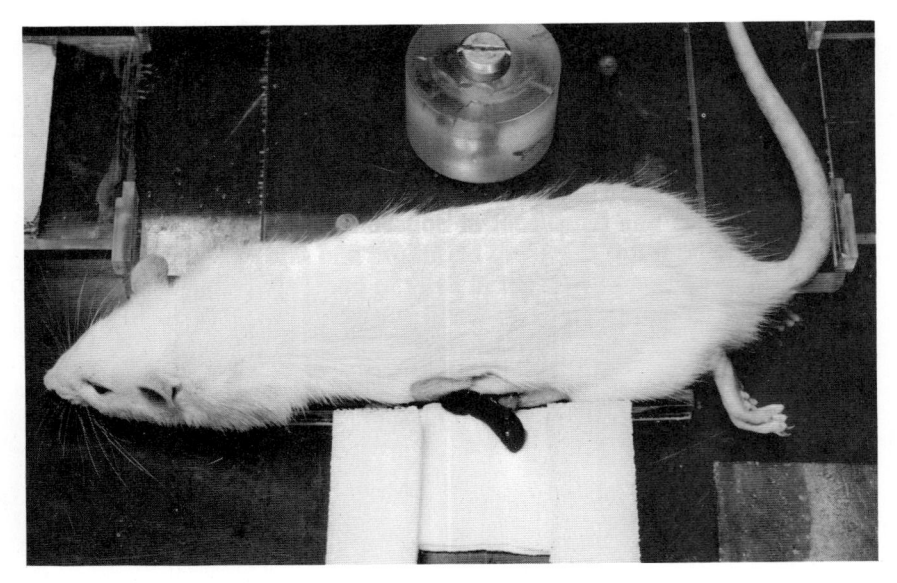

Fig. 4-1. Spleen exteriorized and placed on lead shield

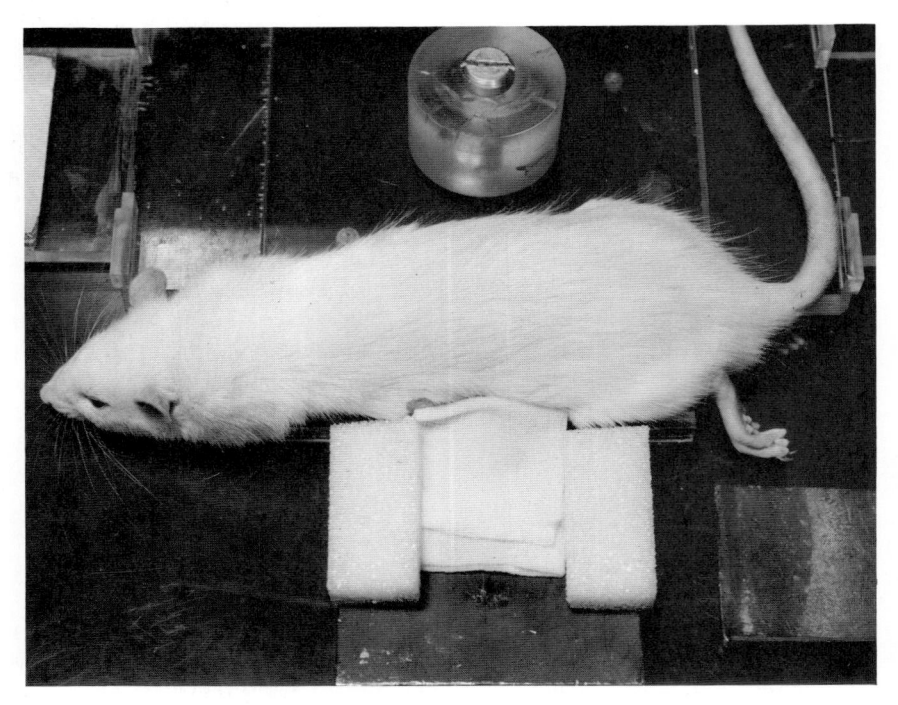

Fig. 4-2. Exteriorized spleen protected by saline-saturated gauze.

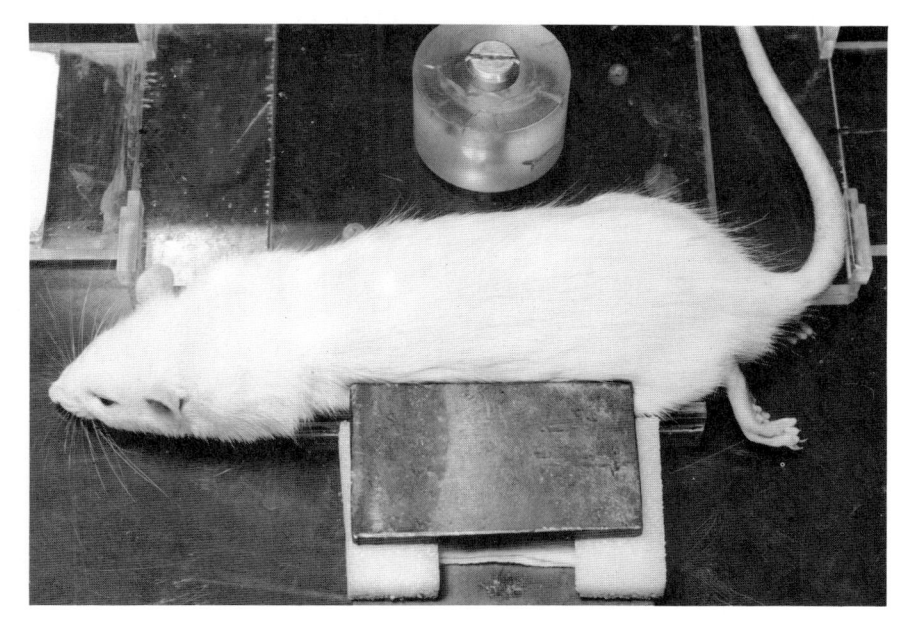

Fig. 4-3. Lead shielding of spleen completed.

The converse technique of shielding the animal's body while irradiating the spleen can, of course, be done exactly the same way except that the shielding must be reversed. The thickness of lead used in the shielding is $\frac{1}{4}$ inch and is about 90% effective for 250 kv X rays.

Special grafting techniques, developed by WHEELER et al. (1966), involve exteriorizing the kidney (Fig. 4.4) and using it as a grafting bed for various other tissues, such as the spleen. The technique is depicted in Figures 4.4 through 4.7. It is useful for transplantation grafts and can be manipulated experimentally by grafting spleen, lymph nodes, bone marrow, and other tissues. Combinations of this technique, X irradiation of organs, and whole-body irradiation could be used to further manipulate biological systems experimentally.

Because of the difficulty of whole-organ grafts owing to vascular connections that are often impossible to reproduce, whole-organ transplants are not technically possible on a large scale. This technique of transplantation allows a good survival and retention of basic morphology of the basic organ for later study. Because of the highly vascular nature of the kidney cortex, it functions as an extremely favorable site for the grafting of slices of other organs. Necrosis of the slices is kept to a minimum if the graft slice is kept to a thickness of 1 to 2 mm.

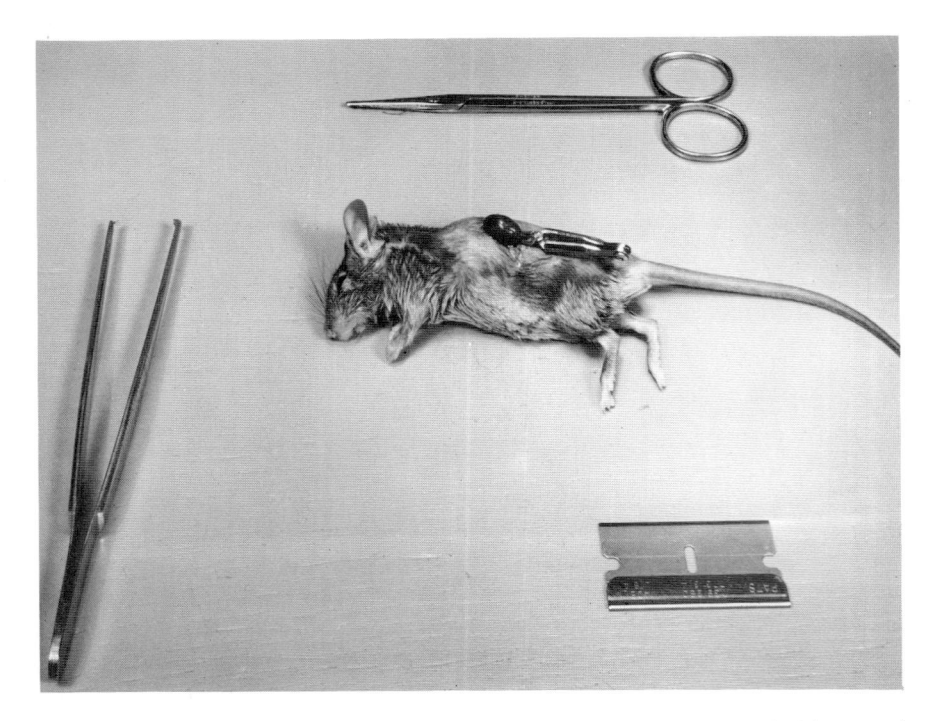

Fig. 4-4. Kidney exteriorized through dorsal lateral incision; renal arteries clamped with an atraumatic hemastat.

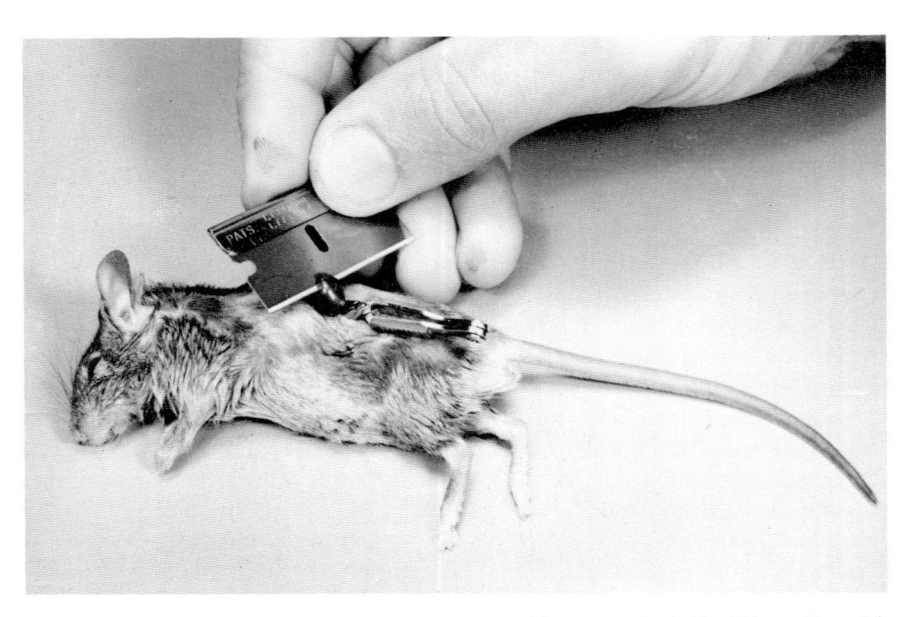

Fig. 4-5. Graft site prepared by removal of thin kidney slice with scalpel or razor blade.

Special Techniques 135

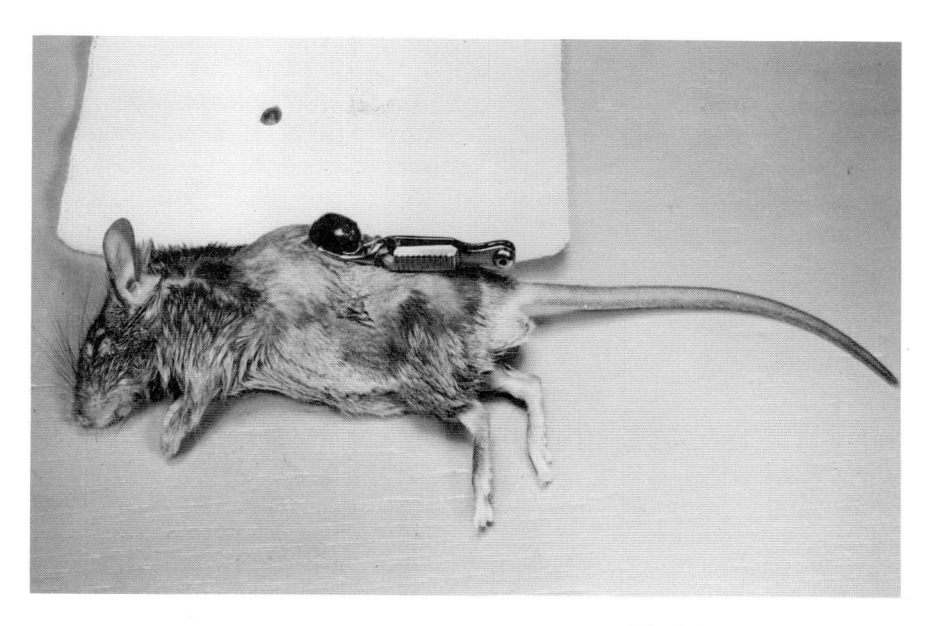

Fig. 4-6. Prepared graft site.

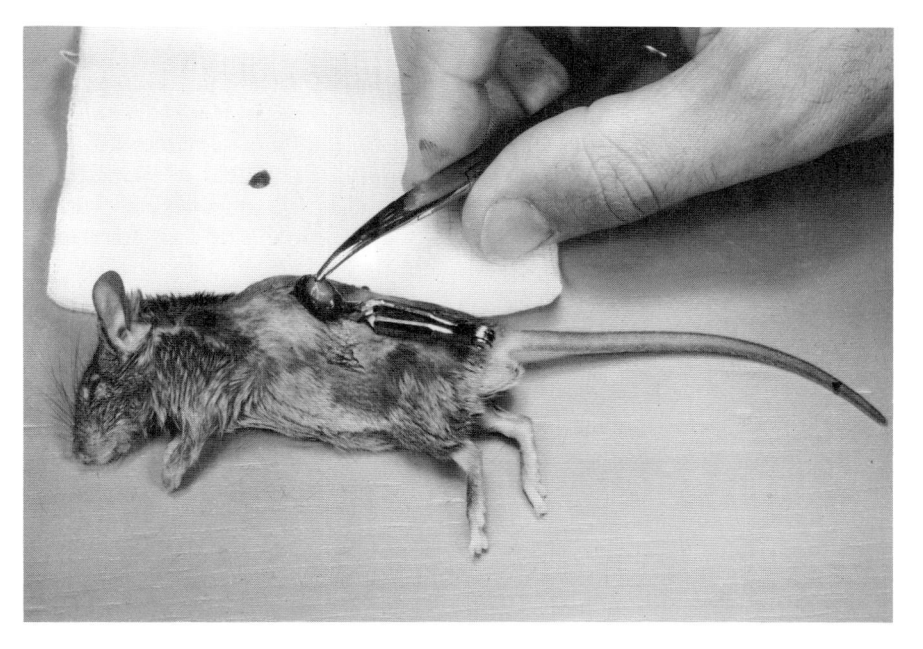

Fig. 4-7. Placing of lymphatic tissue graft on kidney graft bed.

The pedicle of the kidney is clamped gently with an atraumatic vascular clamp, thus minimizing the bleeding from the cut surface of the kidney. WHEELER et al. (1966) point out that for best results it is essential that the cut surface of the donor graft be placed directly on the cut surface of the kidney bed. The graft is then pressed firmly against the kidney bed and held for a few seconds. If the kidney is then placed gently into position, no suturing is necessary for the graft to adhere.

We think this technique has much merit and permits study of healthy, well-vascularized and morphologically intact tissue transplants, at least in inbred mice, and may well be of considerable interest experimentally in inbred and F_1 manipulations in transplantation and secondary disease studies.

DRUG ADMINISTRATION AND DOSAGE

In general, all the techniques described for fluid administration apply to drug administration as well. Of some interest in drug administration is a recent paper reporting a lymphocytopenia in mice following therapy for external parasites with the drug tetraethylthiuram monosulphide (KEAST and COALES, 1967). The paper points out that the drug causes a specific reduction of lymphocytes when used as a treatment for ectoparasites. Even though the lymphocytopenia was apparently transient, it is important to some studies at least to be very cautious in treating experimental animals. However, most therapeutic agents do have modifying effects on the host animal, as well as on the target organisms, and this factor must never be overlooked.

The handbook *Drug Dosage in Laboratory Animals* (BARNES and ELTHERINGTON, 1964) is of value in acquiring a large number of exotic drug dosages in laboratory mice. The drugs listed range alphabetically from acetaldehyde to zoxazolamine, with narcotics, tranquilizers, and hormones in between. The handbook consists of a well-organized bibliography of experimental papers on these drugs. In addition to being a valuable source of information on drugs and drug dosages catalogued as to route of injection, the bibliography is extremely useful in looking up experimental data on specific drugs of interest and their actions and results in a variety of animals, including mice, rats, guinea pigs, cats, dogs, and monkeys. The drugs are indexed as are pharmacological actions and animals. One section includes a comprehensive chart on hormone maintenance and replacement dosages.

Hormone preparations are sometimes used in mice to regulate their breeding cycles for artificial insemination techniques or for timing matings more precisely for caesarean-derivation procedures. Estrus can be induced by subcutaneous injections of pregnant mare serum, followed in 40 hr by human chorionic gonadotropin (HCG). Estrus usually begins 7 to 11 hr after the HCG injection is given, and ovulation occurs shortly thereafter (WOLFE, 1967). WOLFE also describes the techniques for artificial insemination in the mouse in detail in the same paper (1967).

ANESTHETICS

Anesthetic agents have been utilized for many years with a single purpose in mind, that being to prevent the feeling of pain. A more successful recovery can be expected from surgical procedures when the effect of shock owing to severe pain has been reduced by proper anesthesia. Proper anesthesia means a sufficient dose to induce and maintain freedom from pain throughout the entire duration of the procedure. In this section, we limit out discussion to general anesthesia, which will be divided mainly into injectable and inhalation anesthetics. In addition to the various agents in each class, combinations of the agents in the two classes, or combinations of injectable anesthetics, have been used successfully. When using combinations, one agent is usually used to induce while the other is used to maintain. In this manner, the required dose level of highly toxic anesthetics may be reduced.

Injectable Anesthetics

The injectable agents provide a prolonged period of anesthesia without repeated exposure to the agent. Following the injection of nonvolatile agents, controlling the depth of anesthesia is no longer possible; therefore, the actual dose administered is critical. When the calculated dose has been properly administered, injectable anesthetics provide ideal surgical anesthesia for prolonged periods. In addition to the lack of control over the plane of anesthesia, injectable anesthetics cause prolonged recovery periods. In the case of mice, it is usually not a problem. To prevent shock from cold stress during recovery, mice should be placed under warming lamps until fully recovered. Injectable anesthetics, being nonvolatile, are much safer in most laboratory situations than many of the volatile, inhalation anesthetic agents. In this class, anesthetic agents are administered by one of the following methods: (1) per os; (2) subcutaneously (SC); (3) intramuscularly (IM); (4) intraperitoneally (IP); (5) intrathoracically; and (6) intravenously (IV). The route of administration is very important when considering the control of depth, speed of onset, and duration of drug activity. Of the above-mentioned methods, intravenous administration allows the greatest amount of control over depth of anesthesia.

Of the injectable anesthetics used with mice, barbiturates are probably the most common. Although the selection of a general anesthetic should be determined by the experimental design and the animal used, many experiments are designed with very little thought concerning the choice of anesthetic. The effects of each agent are quite varied and must be given careful consideration, especially when various species or strains are being used. The experimental procedure may also benefit from another choice of agents. For example, chloral hydrate can be administered for general anesthesia and is especially useful if electroencephalograms are to be recorded. Barbiturates can alter the EEG pattern; therefore, an alternate choice should be made in such a case. Careful consideration of side effects that may alter an experimental outcome is very important.

In the class of barbiturates, we discuss both long- and short-acting agents, using only generic names and emphasizing only the most common anesthetics in the long- and short-acting groups. Others within each group are listed.

LONG-ACTING BARBITURATES

Pentobarbital. Pentobarbital sodium is the most common drug used in this class. Using a 10% solution administered IP, we find that 0.4 to 0.5 cm^3 is effective. This dose will usually provide surgical anesthesia in adult mice (20 to 35 g) in 5 to 10 min with a duration of 20 to 40 min.

The IV dose of 35 mg/kg and the IP dose of 60 mg/kg have also been reported (BARNES and ELTHERINGTON, 1964, p. 168). Various roentgenographic contrast media have been described that prolong pentobarbital anesthesia in rats (LASSER et al., 1964). Pentobarbital has also been reported to cause tachycardia as well as causing a depression of cardiovascular and spinal cord reflexes (BARNES and ELTHERINGTON, 1964).

SHORT-ACTING BARBITURATES

Thiopental. With thiopental, the onset of anesthesia is about the same as with pentobarbital, but the duration of anesthesia is much shorter. The recovery period of thiopental is also somewhat shorter; however, this fact is usually not significant.

The IV dose of 25 mg/kg has been used with thiopental.

Hexobarbital. This short-acting barbiturate has been used in place of thiopental because it is less likely to produce apnea (LANE-PETTER et al., 1967).

The IV dose is 47 mg/kg (BUSH et al., 1953), the IP dose being 75 mg/kg, and a SC dose of 150 mg/kg has been reported (HEUBNER and SCHULLER, 1936).

Other Short-Acting Agents. Others include methohexital and thialbarbitone. Neither has any particular advantage over the other short-acting barbiturates.

NARCOTICS

Although using or purchasing narcotic drugs requires a special tax stamp and precise records, several narcotic agents have been used in laboratory animals.

Morphine. Morphine, commonly used as a preanesthetic sedative, has been administered to mice at a dose of 2.3 mg/kg IP and 7 mg/kg SC. Morphine also acts as a respiratory depressant, which must be watched carefully.

Meperidene Hydrochloride (Demerol). This particular agent has been used frequently in mice. A synthetic form of morphine, it acts as a depressant of the central nervous system. It can be used as an analgesic but is actually less potent than morphine.

OTHER ANESTHETICS

Chloral Hydrate. The dosage reported for mice is 400 mg/kg. As mentioned previously, chloral hydrate may be used for general anesthesia or for narcosis but must be administered intravenously or orally owing to its irritating effect on perivascular tissue.

Tribromethanol. This drug has been used on occasion in place of pentobarbital. Repeated usage with the same animals will eventually cause liver damage. The IV dose has been reported as 120 mg/kg (HEUBNER and SCHULLER, 1936), although it is best to administer this drug IP, with the reported dose around 250 mg/kg (WOLF and VON HAXTHAUSEN, 1960). Induction following IP administration occurs in approximately 8 to 10 min with a duration of surgical anesthesia lasting about 25 to 30 min.

Ethyl Carbamate (Urethane). For acute or short-term experiments, urethane provides general anesthesia for very long periods of time. However, urethane has been reported to be carcinogenic and, therefore, should not be used repeatedly in long-term experiments. In a current experiment, using urethane as a carcinogen, an IP dose of 1 g/kg provided satisfactory anesthesia in a very short period of time. Recovery at this dose was uneventful; however, an IP dose of 4 g/kg killed all the mice within a 24-hr period. With an IP dose of 2 g/kg, the mice were visibly sick 24 hr after the injection. The sick mice recovered and no early deaths were recorded as a result of the dose used. The minimum lethal dose of urethane has been reported as 2150 mg/kg (FRANKLIN, 1931). A dose level of less than 1 g/kg would most likely be effective for general anesthesia; however, in our tumor induction work, it was not attempted. Because of its carcinogenic effect, urethane must be handled carefully to avoid repeated exposure to the skin. It also produces delayed liver damage with repeated usage.

Inhalation Anesthetics

Inhalation anesthesia refers to the achievement of a general state of anesthesia by a gaseous anesthetic applied by respiration. Inhalation anesthetics are probably used more often with mice than injectable agents because of the simplicity of handling and administration. With inhalation anesthetics, anesthesia is quickly induced, the depth of anesthesia is controllable, and there are few associated side effects. In larger animals, administration of the gaseous agents usually requires constant attention; whereas, with mice, one person can easily handle one or many by using a conventional bell jar. A cotton pledget or gauze squares, saturated with the anesthetic, is placed below a false bottom in the jar. The mouse is placed in the jar and observed until the desired depth of anesthesia is reached, then removed. Upon the mouse's removal from the jar, a nose cone may be used for continued anesthesia, inasmuch as most inhalation agents have a short duration, being exhaled rapidly.

As indicated, the presently available inhalation agents are volatile. A closed system reduces the hazard of fire and explosion; however, with the bell-jar method described, vapor layers will accumulate just above the floor level. Laboratories utilizing these agents must be well ventilated, provide fireproof storage cabinets, and guard against any source of flame while these agents are in use.

A very safe closed system for laboratory animals has recently been reported (SIMMONS and SMITH, 1968). The system was designed primarily for induction and maintenance of mice; however, other small species are easily anesthetized in the closed chamber. Prolonged maintenance of mice is possible by re-entry into the chamber or by placing the animal's nose into the small opening of the lateral tube (Figs. 4.8 and 4.9).

ETHER

Ether, although very flammable, is still one of the most common inhalation anesthetic agents used today. With mice, ether is normally used in the bell-jar system previously described. A high concentration of ether vapor is required to induce anesthesia. It has a strong smell and, in a closed jar system, mice will struggle during the induction period. Mixtures of anesthetic ether with air or oxygen are explosive; therefore, it should not be used in the presence of an open flame.

Ether should be stored in flameproof cabinets or well-vented hoods and not in refrigerators. If excess ether is disposed of by pouring it down the

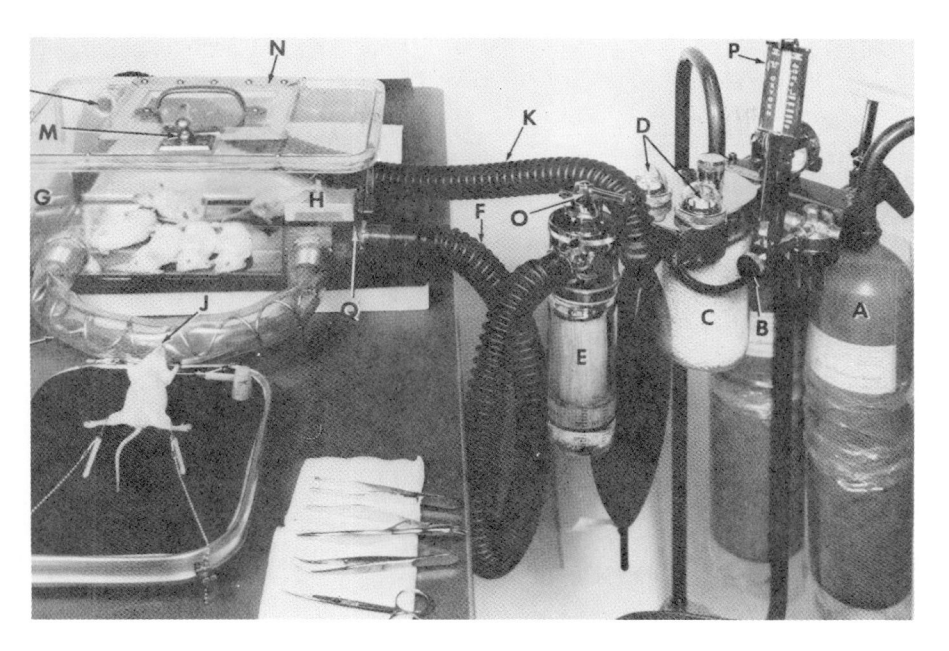

Fig. 4-8. Anesthetic unit for small mammals. A. O_2 tank. B. O_2 control valve. C. Soda lime container. D. Check valve. E. Anesthetic container and vaporizer. F. Inlet tube. G. Polycarbonate chamber (converted mouse cage); 5 inches high, 13 inches long, $8\frac{1}{2}$ inches deep; top is $\frac{1}{4}$-inch Lucite screwed into cage top edge; cutout in top is 4 × 5 inches covered by a 6 × 7-inch hinged piece of Lucite with self-sealing sponge gasket. Volume of chamber plus external circuit (I) is about 8000 cm³. H. Screen-covered curved metal sheet for diverting a portion of the mixture into the external circuit (I). I. External circuit; $1\frac{1}{4}$-inch Tygon tubing containing No. 10 stainless steel coil to maintain patency and rigidity of tubing. Connector fittings are 1-inch stainless steel threaded pipe, with internal and external threaded retaining collars. J. Hole in external circuit for animal's muzzle; stoppered when not in use. K. Outlet tube. L. Exit port for anesthetic and gases; $\frac{5}{8}$-inch stainless steel threaded pipe with internal and external threaded retaining collars. M. Cover latch. N. Cover hinge. O. Vaporizer control valve. P. O_2 flow-rate indicator. Q. Entry port (as in L).

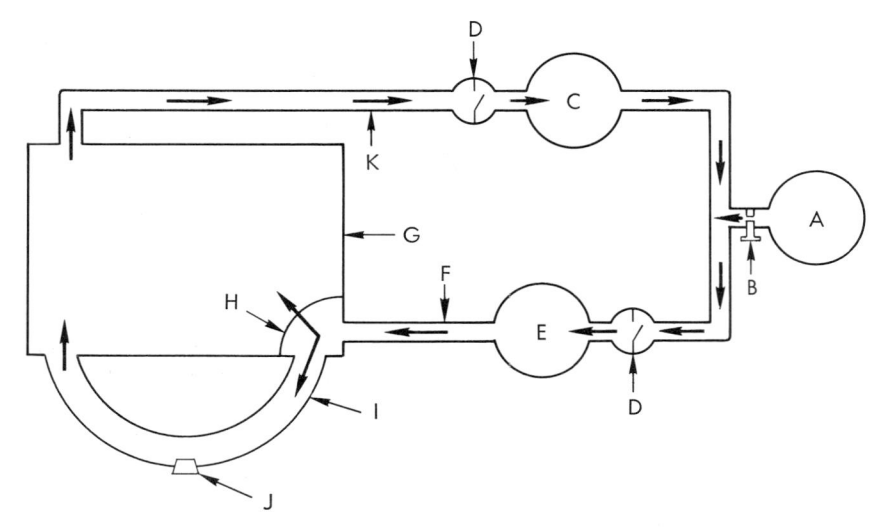

Fig. 4-9. Diagram of closed anesthetic unit. A. Compressed O_2. B. O_2 regulator valve and flow-rate indicator. C. Soda lime container. D. Check valve. E. Anesthetic container–vaporizer. F. Inlet tubing. G. Animal chamber. H. Metal deflector. I. External circuit. J. Stoppered muzzle hole. K. Outlet tubing.

drain, the escaping vapors may flow back into an adjacent lab through the drain system, causing an explosion.

METHOXYFLURANE (METOFANE)

Originally developed as a human anesthetic, Metofane (Pitman-Moore Drug Co.) brand of methoxyflurane is finding widespread use with laboratory animals. A distinct advantage of Metofane in most laboratory situations is its nonexplosive property. It is nonflammable at normal room temperature. As with ether, Metofane works well in a closed or semiclosed system, such as a bell jar. Induction with Metofane in a semiclosed system requires approximately 60 to 90 sec. The muscle-relaxing and analgesic properties of this agent are very good, and because of its low toxicity, overdosage and respiratory failure are usually not problems.

Methoxyflurane will cross the placental barrier and cause depression of near-term fetuses. When Metofane is being used for hysterectomy-derived germfree mice, surgery should be completed as rapidly as possible. Methoxyflurane has also been reported to decrease the total volume of collectable blood in total bleed-out procedures on mice (BRINKMAN and BURCH, 1964).

HALOTHANE

Halothane, although not used much with mice, has several distinct advantages over ether. It is nonirritating, has a pleasant odor, is much more potent than ether, and is not inflammable. Because it is expensive, it is usually used in a closed or semiclosed system. Induction is usually smooth and rapid. HAGEN and HAGEN (1964) have reported problems with maintaining surgical anesthesia in mice when using halothane in a semiclosed system.

CHLOROFORM

Chloroform is being included specifically as an agent *not* to be used. This drug causes damage to the myocardium, liver, and kidneys. In certain strains of mice, the males will die suddenly from nephritis as a result of indirect exposure to small concentrations of chloroform. HEWITT (1956) reported on renal necrosis in mice following accidental exposure to chloroform.

Other Types

ELECTROANESTHESIA

This type of anesthesia with mice has not been successfully developed. Exact placement of the electrodes, in order to pass an alternating current across the brain, is the first problem; restraint and relaxation are additional problems. If the technique could be successfully developed in mice, electroanesthesia would be an asset in certain experiments because no chemical agents are involved. With this form of anesthesia, the operator has fingertip control of depth of anesthesia, and recovery is very fast once the current has been disconnected.

EUTHANASIA

Many methods of euthanasia are available for use on animals that no longer are needed or that have had the experimental procedure completed. The method of euthanasia employed obviously depends a great deal on the number of animals involved. If there are only a few, it is fast and humane to use the method of cervical spinal cord separation. This technique is done by placing the thumb and forefinger just behind the head and applying quick pressure while simultaneously pulling the tail sharply, thus separating the spinal cord completely and causing instantaneous death.

Other methods are ether, barbiturates, and carbon dioxide. A carbon dioxide tank and a plastic bag of any size (depending on the number of animals) are used in the method we prefer. Because carbon dioxide is itself an anesthetic agent, the death is painless and nontraumatic, as well as being relatively inexpensive.

ROENTGENOGRAPHIC TECHNIQUE; RADIOISOTOPE APPLICATION

In relation to the laboratory mouse, radiographs are utilized almost entirely to determine changes resulting from experimental manipulation. The exact technique selected for radiographic work varies considerably owing to different machines and various types of X-ray film. In general, the technique used is included in each publication.

For the detail required when radiographing mice, nonscreen film is recommended. This film is predominantly sensitive to the direct action of X rays and less sensitive to light; therefore, intensifying screens (Cassette-screen radiography) are not required when using this film. Use of this film (as well as ultra-fine detail industrial film) requires 4 to 10 times the milliamperage-seconds (mas) as screen-type radiography.

Several brands of plastic protected film are available (Appendix). Disposable packages permit ease of handling of such film, although it is approximately twice the cost of screen film. For maximum contrast, nonscreen film requires a longer development time and, because of a thicker emulsion, a thorough wash is required between the developer and fixer.

When higher milliamperage is being used, restraint of the animal becomes a major concern. Mice under anesthesia can be positioned and held securely for a sufficient period of time. A thin piece of plastic, with the mouse carefully taped in proper position, works well. The plastic is placed on top of the film, which may be subdivided by a lead rubber shield cut to protect three-quarters of the film, especially if using 8×10 or larger.

According to a recent report (PYKE, 1966), fine-grain X-ray film, such as Kodak AA, will produce the best results when radiographing in the range of 30 to 80 kvp. X rays of lower kilovoltage will produce better radiographs with higher contrast and more detail than those using higher kilovoltage. In a recent report (HOPKINS et al., 1966), strontium-90-induced skeletal changes were examined in rats by using the following exposure factors: 0.1 sec exposure, 100 ma and target-to-skin distance of 1.7 m. In this particular study, screen radiography was employed, and Kodak Blue Brand film was used in combination with Patterson Par–Speed intensifying screens.

FINKEL et al. (1961), in a publication on strontium-induced osteosarcomas in mice, reported exposure technique of 0.75 sec, 30 ma, 32 kv, and

19 inches focal film distance. In a more recent publication (FINKEL et al., 1966), on virus-induced osteosarcomas, the exposure technique was described for a rotating anode X-ray tube with 0.5 mm and 1.5 mm focal spots. Kilovoltage varied with the size of the mouse, 15 kvp being used for newborn mice and 35 kvp for larger mice. Focal-film distance varied from 12.5 to 15 inches; exposure times varied from 0.15 to 1 sec. Living mice were radiographed at 200 ma by using the larger focal spot, whereas dead mice were radiographed with 50 ma by using the finer focal spot (0.5 mm). For this type of work, Kodak Type AA industrial film is reported to be far superior to standard nonscreen X-ray film. With the above technique, the first bone tumor was observed radiographically 21 days after virus injection as a small swelling of the proximal end of the eleventh rib.

In addition to nonscreen, Type M, and Type AA industrial film, dental films may be of practical value for radiographing specific parts, such as the lungs of mice. The authors have attempted to delineate lung neoplasms by using special techniques, such as angiography or bronchiograms. The early data from these methods have been inconclusive. However, MARGULIS et al. (1964) reported on a technique that they used in rats in an attempt to delineate simulated abdominal tumors. Radiographic examination of the abdomen and the abdominal organs (celiography) of rats by using various contrast mediums is discussed. According to this report, iothalamic acid produced the best overall results. Other contrast media evaluated included Hypaque, Renografin, barium sulfate, and air.

Lymphangiography and angiographic techniques have been utilized to study various postirradiation effects. Angiographic procedures have been described by MARGULIS et al. (1961). NOONAN et al. (1967), using angiographic techniques, have described the specific evaluation of the effect of trauma and irradiation on transplanted lymphosarcomas of mice.

Using Thorotrast, MCALISTER and MARGULIS (1963) have reported on the application of angiography to study malignant tumors in mice following irradiation.

A recent and welcome addition to the field of laboratory animal radiography is a book entitled *Roentgen Techniques in Laboratory Animals*, ed. by Benjamin Felson, M.D. (Philadelphia: W. B. Saunders Company, 1968), 245 pp.

Radioisotopes

Many isotope-labeled compounds are presently available commercially. The applications and usefulness of radioisotopes in relation to experiments involving mice are extensive. In this text, we include only a reference list relating to the subject.

Recommended Reading on Radioisotope Applications

Boyd, G. A. *Autoradiography in Biology and Medicine* (New York: Academic Press, 1955), 399 pp.

Gude, William D. *Autoradiographic Techniques: Localization of Radioisotopes in Biological Material* (Englewood Cliffs, N.J.: Prentice-Hall, Inc., 1968), 113 pp.

Hermias, Sister Mary, and Sister Mary Joecile. *Radioactivity: Fundamentals and Experiments* (New York: Holt, Rinehart & Winston, Inc., 1963), 209 pp.

Hughes, W. L., V. P. Bond, G. Brecher, E. P. Cronkite, R. B. Painter, H. Quastler, and F. G. Sherman. "Cellular Proliferation in the Mouse as Revealed by Autoradiography with Tritiated Thymidine." *Proc. Natl. Acad. Sci.* **44**: 476–483, 1958.

Hurlburt, Evelyn M. *Radioisotope Technique for Instruction in the Biological Sciences. A List of Annotated References* (U.S. Atomic Energy Commission TID–21262, 1964). (Available from Clearing House for Federal and Technical Information, National Bureau of Standards, U.S. Dept. of Commerce, Springfield, Va.)

Johnson, Horton A., and Eugene P. Cronkite. "The Effect of Tritiated Thymidine Mortality and Tumor Incidence in Mice." *Radiation Res.* **30**: 488–496, 1967.

Laboratory Experiments with Radioisotopes for High School Science Demonstrations, ed. by Samuel Schenberg, Superintendent of Documents (Washington, D.C.: U.S. Atomic Energy Commission, 1958).

Quastler, H., and F. G. Sherman. "Cell Population Kinetics in Intestinal Epithelium of the Mouse." *Exp. Cell Res.* **17**: 420–438, 1959.

Rogers, Andrew W., ed. *Techniques of Autoradiography* (Amsterdam: Elsevier Publishing Company, 1967), 335 pp.

Special Sources of Information on Isotopes. U.S. Atomic Energy Commission TID–4563, 3rd Rev., Division of Isotopes Development, Washington, D.C. (Obtainable from Division Technical Information Extension, U.S. Atomic Energy Commission, P.O. Box 62, Oak Ridge, Tenn. 37830.)

Tonna, E. A., and E. P. Cronkite. "Autoradiographic Study of Periosteal Cell Proliferation with Tritiated Thymidine." *Lab. Invest.* **11**: 455–462, 1962.

CHAPTER 4 REFERENCES

Barnes, C. D., and L. G. Eltherington. *Drug Dosage in Laboratory Animals: A Handbook* (Berkeley, Calif.: University of California Press, 1964), 302 pp.

Brinkman, D. C., and G. R. Burch. "Metofane Brand of Methoxyflurane Anesthesia in a Variety of Animal Species." *The Allied Veterinarian*, March–April, 1964.

Bush, M. T., T. C. Butler, and H. L. Dickison. "The Metabolic Fate of Evipal (Hexobarbital) and of 'Nor Evipal.'" *J. Pharmacol. Exp. Therap.* **108**: 104–111, 1953.

Ersek, Robert A. "Murine Splenectomy." *Am. J. Vet. Res.* **29**: 755–758, 1968.

Finkel, Miriam P., Patricia J. Bergstrand, and Birute O. Biskis. "The Latent Period, Incidence, and Growth of Sr^{90}-Induced Osteosarcomas in CF1 and CGA Mice." *Radiology* **77**: 269–281, 1961.

Finkel, M. P., P. B. Jinkins, J. Tolle, and B. O. Biskis. "Serial Radiography of Virus-Induced Osteosarcomas in Mice." *Radiology* **87**: 333–339, 1966.

Franklin, K. J. "The Pharmacology of Some Compounds Allied to Chloral and to Urethane." *J. Pharmacol. Exp. Therap.* **42**: 1–7, 1931.

Hagen, E. O., and J. M. Hagen. "A Method of Inhalation Anesthesia for Laboratory Mice." *Lab. Animal Care* **14**: 13–17, 1964.

Heubner, W., and J. Schuller, eds. *Narcotics of the Aliphatic Series; Handbook of Experimental Pharmacology (Heffter)*, Vol. 2 (Berlin: Springer Verlag, 1936), 283 pp. (In German.)

Hewitt, H. B. "Renal Necrosis in Mice After Accidental Exposure to Chloroform." *Brit. J. Exp. Pathol.* **37**: 32, 1956.

Hopkins, B. J., G. W. Casarett, R. C. Baxter, and L. W. Tuttle. "A Roentgenographic Study of Terminal Pathological Changes in Skeletons of Strontium-90 Treated Rats." *Radiation Res.* **29**: 39–49, 1966.

Jacobson, Leon O., E. K. Marks, E. O. Gaston, M. Robson, and R. E. Zirkle. "The Role of the Spleen in Radiation Injury." *Proc. Soc. Exp. Biol. Med.* **70**: 740–742, 1949.

Keast, D., and M. F. Coales. "Lymphocytopenia Induced in a Strain of Laboratory Mice by Agents Commonly Used in Treatment of Ectoparasites." *Australian J. Exp. Biol. Med. Sci.* **45**: 645–650, 1967.

Lane-Petter, W., A. N. Worden, B. F. Hill, J. S. Paterson, H. G. Vevers, and the Staff of UFAW, eds. *The UFAW Handbook on the Care and Management of Laboratory Animals*, 3rd ed. (Baltimore: The Williams & Wilkins Co., 1967), 1015 pp.

Lasser, E. C., G. Elizondo-Martel, and R. C. Granke. "The Roentgen Contrast Media Potentiation of Nembutal Anesthesia in Rats." *Am. J. Roentgenol. Radium Therapy Nucl. Med.* **91**: 453–460, 1964.

Margulis, A. R., H. J. Burhenne, and O. N. Rambo. "Evaluation of Celiography in Rats." *Radiology* **82**: 290–295, 1964.

Margulis, A. R., E. Carlsson, and W. H. McAlister. "Angiography of Malignant Tumors in Mice." *Acta Radiol.* **56**: 179–192, 1961.

McAlister, W. H., and A. R. Margulis. "Angiography of Malignant Tumors in Mice Following Irradiation." *Radiology* **81**: 664–675, 1963.

Munoz, J. "Production in Mice of Large Volumes of Ascites Fluid Containing Antibodies." *Proc. Soc. Exp. Biol. Med.* **95**: 757–759, 1957.

Noonan, C. D., A. R. Margulis, H. M. Patt, and J. Stoughton. "Angiographic Evaluation of the Effect of Trauma and Irradiation on Transplanted Lymphosarcoma in Mice." *Radiology* **89**: 923–928, 1967.

Pyke, Ralph E. "A Method of Measuring the Rate of Skeletal Maturation of the Rat from Radiographs of the Forepaw." *Lab. Animal Care* **16**: 393–394, 1966.

Riley, Vernon. "Adaptation of Orbital Bleeding Technique to Rapid Serial Blood Studies." *Proc. Soc. Exp. Biol. Med.* **104**: 751–754, 1960.

Shrewsbury, Marvin M. "Rate of Flow and Cell Count of Thoracic Duct Lymph in the Mouse." *Proc. Soc. Exp. Biol. Med.* **99**: 53–54, 1958.

Simmons, M. L., and L. H. Smith. "An Anesthetic Unit for Small Laboratory Animals." *J. Appl. Physiol.* **25**: 324–325, 1968.

Wheeler, H. Brownell, Joseph M. Corson, and Gustave J. Dammin. "Transplantation of Tissue Slice in Mice." *Ann. N.Y. Acad. Sci.* **129**: 118–129, 1966.

Wolf, A., and E. F. von Haxthausen. "Toward the Analysis of the Effects of Some Centrally-Acting Sedative Substances." *Arzneimittel-Forsch.* **10**: 50–52, 1960.

Wolfe, H. Glen. "Artificial Insemination of the Laboratory Mouse (*Mus musculus*)." *Lab. Animal Care* **17**: 426–432, 1967.

Technological Advances in Laboratory Animal Management

FIVE

INTRODUCTION

In recent years, many developments have taken place in the field of laboratory animal care. Technical methods for achieving practical purposes have evolved as a result of the effort required to handle the large numbers of mice involved in the many long-term experiments utilizing these animals. For example, computerized breeding and experimental programs are capable of handling the large volumes of data necessary for running production colonies or for analyzing experimental results. Information of this type can be stored indefinitely and analyzed whenever necessary. This system greatly reduces the complexity of recordkeeping and has proved to be much more accurate than maintaining manual records.

In addition to computerized breeding programs, automated systems have been developed to aid in modern care of large numbers of mice. An example is the automatic watering system, which has eliminated the need for frequent water-bottle changes. The obvious reduction in the amount of manual labor required is an important contribution to an area of concern in mammalian research, that is, reduction of cost.

Filter tops have played a very important role in the concept of pathogen-free animals. In this chapter, we discuss several of the more recent techno-

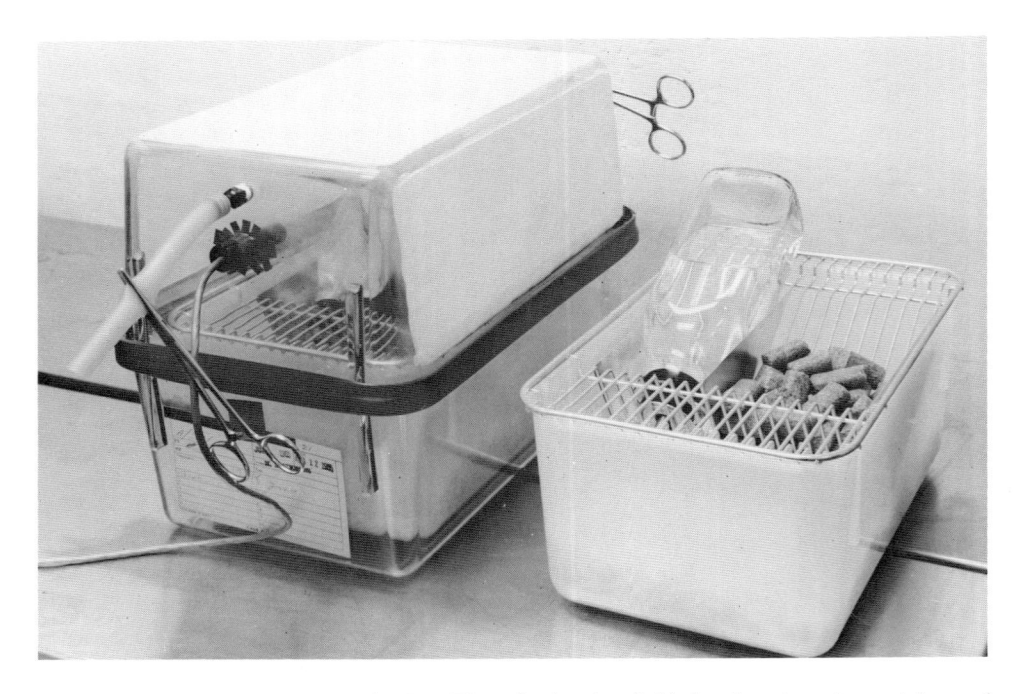

Fig. 5-1. Hypoxia chamber (left) showing air exchange tubes and sensing element. Cage at right shows size comparison.

logical advances associated with the laboratory mouse, including the application of filter tops and use of a hypoxia chamber.

HYPOXIA CHAMBER

In 1966, LANGE et al. reported the development of a hypoxia chamber for the production of polycythemic mice (Fig. 5.1). The dimethyl silicone rubber was hand-fabricated in the laboratories of General Electric Company. Silicone rubber film (0.001-inch thick) was bonded on both sides of the membrane with 0.005-inch-thick "Dacron" mat in order to improve the membrane's durability, protect it from abrasion, and support a pressure differential.

The enclosures were fabricated by modifying two standard "Lexan" animal-cage bottoms. One cage bottom, which was subsequently inverted to become the top of the enclosure, was modified by cutting openings (amounting to 143 sq inches) in the sides and top. The openings were covered with the

Dacron-backed silicone rubber membrane. Two tube fittings were installed, one in each end of the top of the enclosure, for the purpose of flushing the cage with gas mixtures. An additional hole was cut to allow the positioning of a sensor to monitor the oxygen concentration. Cages were opened three times a week for cleaning, and oxygen content was measured by a Beckman polarographic oxygen analyzer.

In a series of experiments, different numbers of animals were allowed to reach oxygen equilibrium without flushing the cages. An average of 470.5 min (ranging from 442 to 520 min) was required for equilibrium to be reached. The subsequent oxygen concentration was found to be dependent on the number of animals in the enclosure (Table 5.1). With 11 animals in the enclosure, the range of carbon dioxide determinations made on air inside the enclosure was 1.5 to 2.1%.

TABLE 5.1

Effect of Number of Mice on Oxygen Content in Cages Covered with Permselective Membrane

Number of mice in enclosure	Approximate O_2 content (%) (24-hr average)	Approximate equivalent altitude (ft)
1	18.2	3900
3	15.0	8800
5	11.8	15,000
7	9.3	21,000
9	8.1	24,000
11	6.8	28,000

When 11 animals were placed in an enclosure for a period of 3 weeks, there was a progressive rise from 47.65 to 73.5% in hematocrit values. Eight days after the animals were removed from the enclosure, the mean hematocrit was 56.89%.

Based on the selective permeability of certain organic materials, the concept of using semipermeable membranes for separation of gases and vapors dates back to the nineteenth century. The term "selective permeability" means that gas A will permeate the membrane faster than gas B, and that gas B will permeate the membrane faster than gas C, and so on. These unique properties led to the choice of the term "permselective membranes," which separates these materials from those that are merely porous and pass gases in a nonselective manner. Some of the properties of the dimethyl silicone rubber membranes and their use in a blood oxygenator have been described by ESMOND and DIBELIUS (1965).

In the enclosures, as the animals consumed oxygen, the partial pressure of oxygen inside the enclosure was lowered below that of the ambient air. This process produced an oxygen partial pressure difference across the membrane into the enclosure according to the following equation:

$$N_{O_2} = \frac{PrO_2}{t} A (PX_{O_2} - px_{O_2})$$

where

$N = $ cm^3/sec (NTP)
$t = $ film thickness in cm
$PrO_2 = $ permeability of O_2
$P = $ pressure of feed gas in cm Hg
$X_{O_2} = $ molecular fraction of O_2 in feed gas
$p = $ pressure of product gas in cm Hg
$x_{O_2} = $ molecular fraction of O_2 in product gas
$A = $ area in cm^2

As the animals produced CO_2, the partial pressure of CO_2 built up inside the enclosure, thus producing a partial pressure difference of CO_2 across the membrane. Carbon dioxide permeated the membrane from within the enclosure, according to the equation:

$$N_{CO_2} = \frac{PrCO_2}{t} A (PX_{CO_2} - px_{CO_2})$$

Because of the difference of permeability between CO_2 and O_2, the partial pressure difference to drive the CO_2 out of the cage is less than one-fifth that necessary to supply the O_2 to the animals. For example, if the partial pressure of O_2 is reduced from 21% outside the cage to 11% inside the cage, the CO_2 concentration inside the cage will rise to about 2%.

The enclosures represent a new method for producing polycythemia in experimental animals. The animals appear healthy and eat well while in the enclosures, but they lose weight and consume less water than do animals kept in room air. The mice regain the weight lost in 4 days after removal; therefore, it is assumed that the weight loss represents primarily fluid loss. Animals held in a hypobaric chamber also lose weight and after removal regain it in a similar manner.

Although these results are concerned with enclosing animals in membrane cages and measuring the effect on red cell parameters, it is obvious that the enclosure would be easily adaptable to other physiological experiments in which a mild-to-moderate hypoxic stimulus is needed.

FILTER-CAGE SYSTEMS

For many years, investigators have been striving to achieve a new level of freedom from infectious disease in laboratory animals. With the development of successful germfree techniques, several species became available to serve as diseasefree breeder stock. Barrier-type holding facilities were then developed rapidly to maintain large numbers of rodents in a diseasefree state. Although both of these methods are highly effective, not all laboratories can afford a germfree operation or a barrier facility.

In a much less expensive, yet equally effective, manner, KRAFT (1958) eliminated epidemic diarrhea of infant mice (EDIM) from a colony by housing them in uniquely designed filter cages after selecting EDIM-free breeder stock. A bacteriological transfer hood was also used when changing cages or handling the mice. These early filter cages were made of galvanized iron cylinders provided with a flat, snugly fitting galvanized cover. The main portion of this cylinder was fashioned of galvanized wire mesh, about which was placed a piece of fiberglass insulating material (Aerocor PF 105). This fiberglass-mesh area was as large as possible to allow sufficient exchange of inside and outside air. KRAFT reported no internal or external evidence of diarrhea in the mice born and housed in filter cages. Also, 97% of the young born were still alive at 18 days of age. This work started the era of filtered cages as a means of controlling disease conditions in laboratory animals.

Since that time, many types of filter media formed into various shapes and sizes have been employed to maintain pathogenfree mice. A few of the more recently developed filter-cage systems are depicted in Fig. 5.2a,b,c. Although filter-cage systems vary considerably in the type of filter media, size, shape, and so forth, they all have several common criteria of effectiveness. The selected media (1) must be capable of at least 98% filtration efficiency of all particles down to 0.3μ; (2) must lend themselves readily to fabrication—that is, be easily cut, sewn, folded, stapled, glued, compressed, and so on; (3) must be made in such a way so as to provide a tight seal around edge of cage; (4) must allow sufficient exchange of air between the inside and outside of the cage; (5) must be of sufficient quality to allow frequent handling and sterilization; (6) should not reduce the efficiency of the cage and/or water bottle changing process by more than 20%; and (7) should not greatly alter the internal environmental cage conditions—that is, temperature, humidity, percent of oxygen, carbon dioxide, nitrogen, and ammonia.

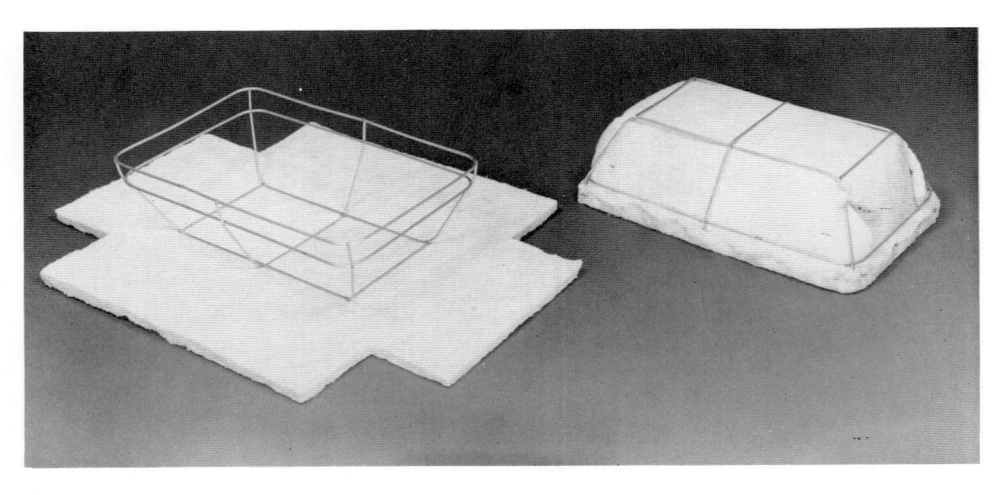

154 *Technological Advances in Laboratory Animal Management*

(a)

Filter-Cage Systems 155

(b)

(c)

Fig. 5-2. (a) Makeup of fiberglass filter top, and filter top in place. (b) Commercially available paper filter top, showing hold-down bar and neoprene gasket around top edge of cage. (c) Dacron filter top, presently under investigation.

Two recent papers (SIMMONS et al., 1967; BRICK et al., 1969) describe in detail pathogenfree operations utilizing the filter top shown in Fig. 5.2a. The filter medium (FM–004, Owens–Corning) is stapled to the inside of the wire frame. The compression of the filter material around the top edge of the cage provides a very tight seal when in proper position. The water bottle and feed are placed on the cage top, under the filter top. Daily care of breeder mice requires the top to be off the cage approximately 45 sec in a 24-hr period. This limited exposure of the mice greatly reduces the possibility of airborne contamination from cage to cage, or from the environment. Even with filter tops, additional precautions are necessary to maintain pathogenfree mice. If the colony is not a pathogenfree breeding colony, all incoming mice must be from a pathogenfree source. In addition, face masks, caps, gloves, and forceps should be used when handling the mice. The feed and bedding may or may not be sterilized, depending upon the type of operation. The water used should be hyperchlorinated or acidified. Room temperature should be controlled for the comfort of the animals and the animal caretakers.

The results of several years of work utilizing filter-top cage systems here at the laboratory adequately demonstrate the practicality of the utilization of such systems. As more commercial breeders supply SPF and GF mice, the investigator will be able to maintain better animals for research purposes at a reasonable cost.

The most recent innovation in filter-cage systems is the "Dacron" filter cap (Fig. 5.2c) designed by Charles Lee Associates. At the present time, extensive testing is being conducted to evaluate the effectiveness of this new filter material. To date, this filter cap has met all the requirements outlined in this section, and therefore may prove to be very useful in the near future.

COMPUTERIZED DATA STORAGE AND RETRIEVAL

One of the most rapidly expanding fields in the area of biomedical research is the storage, retrieval, and analysis of data through computerized programming. Programs are written for many different purposes, of course, but the systems generally have the following objectives in common: (1) simplicity of data entry, (2) prevention of data loss, (3) data retrieval as required, (4) specific comparisons or analysis of data, and (5) supervisory control of various operations using program outputs of stored data.

At the present time, computers are used in a variety of ways (MCKELVIE et al., 1967; TAFT, 1968). In addition to computer programming for breeding colony operations and experimental data, computers have been used for complete data processing in a clinical laboratory (TAFT, 1968). By using an on-line system, a general purpose computer can be attached directly to the analytical instruments. In this manner, the on-line system can automatically read and record test results as they are generated, correlate several tests for the same patient, automatically validate and calculate the clinical values, and prepare a final report for the attending physician. This system also provides constant monitoring of the equipment during any analysis and can alert the technologist immediately if retesting is necessary.

Maintenance of breeding records is a very important and time-consuming part of a laboratory breeding colony. As a result, computer programs have been developed for experimental animal production colonies (KALLMAN, 1966; FRITZ et al., 1967; SERRANO and AMSBURY, 1967). The weekly events in a breeding colony are quite repetitious; therefore, any two programs will differ primarily in their method of input and the form of the program output. A recent publication (FRITZ et al., 1967) describes the use of daily operation and mating worksheets as a method of input. This program provides a continuous analysis of the colony breeding efficiency, as well as a means of early recognition of disease problems. The printout sheets of any program will vary according to the desires of the investigator and the type of data being recorded.

In a more recent publication on computer processing of mouse breeding data (SERRANO and AMSBURY, 1967), a commercially available "Data Collection System" is discussed. This system consists of two interconnected electromechanical devices referred to as (1) a sending station and (2) a variable data input station for manual entry. Both pieces of equipment are connected to a data receiving station located in another building. Data are put into the system by means of 80-column punched cards and badges and the variable data input station. In this particular colony, the sending station and variable input station are both located in the breeder room. The breeder technician in this room is trained to operate the equipment.

A third type of computerized breeding program presently in use is a system that combines both of the previously described methods of input into one program. A single 80-column breeding colony daily input sheet is used routinely, although a centrally located data input station and variable manual entry unit are always available (Fig. 5.3). With this system, the colony operation is not totally dependent upon mechanical input stations or keypunch operators. This breeding program was developed to (1) store all breeding colony information for inbreds or hybrids, (2) provide weekly assignment sheets for animal caretakers, (3) provide summary sheets and graphs of offspring born and weaned, (4) analyze total colony production performance, and (5) store research cage assignments for entry into experimental programs. Additional programming can integrate such a breeding program directly

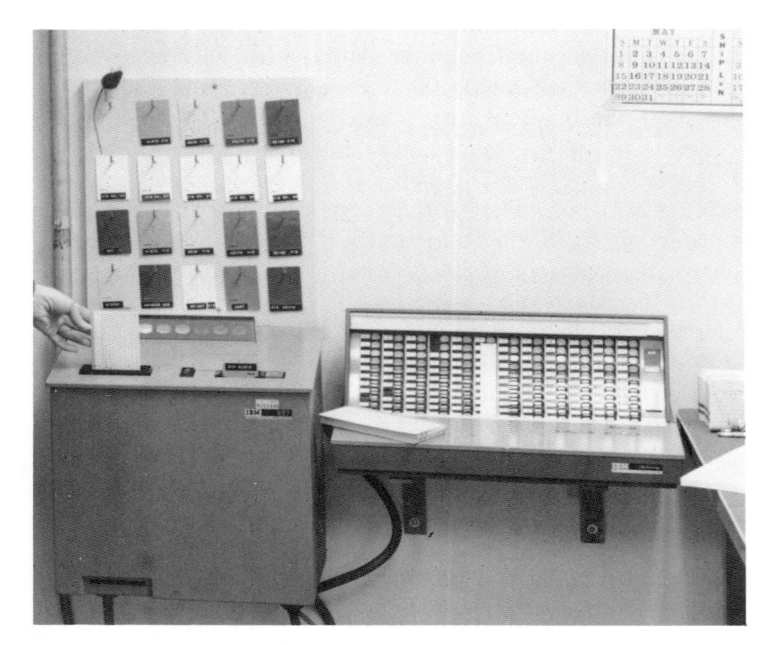

Fig. 5-3. IBM 357 card and badge input station (left), and IBM 372 manual input station, interconnected.

into an experimental program that in itself can be tied directly to a computerized pathology program.

Breeding colony data are stored on a master tape and updated weekly. With the weekly assignment sheet provided by such a program, each breeder technician is capable of handling many more breeder mice. The time and cost for maintaining computer programs are less than those required for manual recordkeeping, whereas reliability and accuracy are greatly improved. In a breeding colony, early recognition of increasing death loss of weanling mice is important. This recognition is possible through the summaries and analysis provided. Part of the weekly output sheet informs the breeder technician when to mate, wean, and retire mice. A computer program of this type also provides a means of storing historical records and allows tracing of the genetic lineage of a colony of inbred or hybrid mice.

Any automatic data collection system must have some method of input—that is, manual input station such as an IBM 357 unit and IBM 372 manual entry located in each breeder room, or one in a central location, or a data input sheet. The direct use of the input sheet eliminates the necessity of personnel making input transactions through a manual input station, which in turn eliminates a possible source of error. The input data sheet must be correct and legible; therefore, some additional time is required by the animal caretaker in the room. It is well compensated for, however, by the time saved in running transactions and filing punched cage cards and badges necessary for sending a transaction through a manual input station.

Manual input stations in each animal room have the additional disadvantages of (1) transaction noise in the breeder room, (2) difficulty of servicing during mechanical failure, (3) cost of multiple units and additional cable, (4) requirement for storage of punched cards and badges in the room, and (5) training of the animal caretakers to make the transactions.

A computerized breeding program is not inexpensive. In making a decision to go on a computer program, one must consider the number of animals involved, the duration and complexity of the colony, the volume of records or data to be maintained, the nearness of a computer center, and the direct supervisory benefits provided by such a program. A computerized program, to be effective, must be flexible to accommodate changes in future operations. When writing a program, it is very important that we are sure the operation or experiment governs the program, and not vice versa.

Programs generating cage cards and maintaining animal inventories as well as receiving reports are advantageous for weekly summary. It is possible to maintain continuity between cage and animal inventory numbers and experimental programming. For example, we can use cards punched with appropriate data on the arrival of mice to make entries on specific experiments by running them through remote control units in laboratories, thus carrying forward all prepared data to the specific experiment number automatically.

AUTOMATIC WATERING SYSTEMS

During recent years, a great many of the routine practices of laboratory animal care have been automated. For example, water bottles and cages that in the past were washed monthly are now usually washed on a regular weekly schedule in automatic or semiautomatic cage and bottle washers. Chlorination of the water may still be done on a manual basis; however, semiautomatic chlorinators are now available for this purpose (Appendix).

Although these factors have not necessarily reduced the labor force, they certainly have created more time for direct attention to the animals by the caretakers. It makes good sense to increase the automation of animal care to extend even to the food and water. The advantages of always having fresh food and water available in small quantities are obvious. Although automatic food control is quite likely a problem to be dealt with in the more distant

160 *Technological Advances in Laboratory Animal Management*

(a)

Fig. 5-4. (a) Cage of mice positioned for preliminary trial of automatic watering system. (b) Automatic timer and regulators for automated watering system.

(b)

future, the problem of automatic watering systems suitable for mice is quite current.

To meet our requirements, an automatic watering system for mice should (1) be completely automatic, (2) prevent cage-to-cage spread of disease, (3) be economically feasible, (4) be convenient for caretakers and investigators to use, and (5) be absolutely reliable (e.g., cages must not be flooded, yet we must have confidence that they have an ample supply of water at all times).

At the time of the writing of this book, we are not satisfied with our present systems. The five criteria just mentioned are difficult to meet, and the greatest difficulty of all is the valve itself. Mice pack pieces of bedding and fecal material into the sipper tube or valve of the watering system, causing automatic valves to stick in the open position, with wet or drowned mice as the result. If the valve is removed from the cage, you then have the problem of smaller mice having difficulty in reaching the water source.

However, all these technical details will be surmounted in time. At present, we feel that we have devised a system that may meet many of our criteria, provided a more suitable valve can be found. It consists of a hard plastic (chlorine-resistant) line with pressure-reduction valves and a line bypass system attached to a timer (Fig. 5.4a,b). This simple system allows us to flush the line for x number of minutes every 30 min, every hour, or every 2 hr, and so forth, at a water pressure considerably higher than normally maintained. By this method we can maintain the chlorine level in the line at any predetermined level. The high-pressure flush clears the lines completely of any debris or particles and does not allow a drift downstream into the water supply of another cage. Work on additional technical details of valves is in progress, and we feel confident that new mouse buildings constructed in the future will include at least the basic piping for a central automatic chlorinated watering system. This system can be installed relatively inexpensively at the time of construction. After construction, of course, installation becomes very expensive.

Other automatic systems will no doubt appear from time to time; however, each one must be carefully evaluated against established criteria of the specific operation, whether it be production, experimental, or toxicological.

CHAPTER 5 REFERENCES

Brick, J. O., R. F. Newell, and D. G. Doherty. "A Barrier System for a Breeding and Experimental Rodent Colony: Description and Operation." *Lab. Animal Care* **19**: 92–97, 1969.

Esmond, W. G., and N. R. Dibelius. "Permselective Ultra-Thin Disposable Silicone Rubber Membrane Blood Oxygenator: Preliminary Report." *Trans. Am. Soc. Artificial Internal Organs* **11**: 325–329, 1965.

Fritz, Thomas E., Merlin H. Dipert, and Robert J. Flynn. "The Utilization of a Digital Computer for Analysis and Management of Rodent Breeding Data." *Lab. Animal Care* **17**: 114–129, 1967.

Kallman, Robert F. "A System for Large-Scale Inbred Mouse Production, and a Computer Method to Aid in Subline Selection." *Lab. Animal Care* **16**: 345–359, 1966.

Kraft, L. M. "Observations on the Control and Natural History of Epidemic Diarrhea of Infant Mice (EDIM)." *Yale J. Biol. Med.* **31**: 121–137, 1958.

Lange, Robert D., M. L. Simmons, and Norman R. Dibelius. "Polycythemic Mice Produced by Hypoxia in Silicone Rubber Membrane Enclosures: A New Technique." *Proc. Soc. Exp. Biol. Med.* **122**: 761–764, 1966.

McKelvie, D. H., T. J. Powell, and Sally Lee. "Programming for Computer Storage, Retrieval, and Analysis of Clinical Chemistry Data in a Life-Span Study on Beagles." *Lab. Animal Care* **17**: 494–500, 1967.

Serrano, Louis J., and Carlene Amsbury. "Semiautomatic Recording and Computer Processing of Mouse Breeding Data." *Lab. Animal Care* **17**: 330–341, 1967.

Simmons, M. L., C. B. Richter, J. A. Franklin, and R. W. Tennant. "Prevention of Infectious Diseases in Experimental Mice." *Proc. Soc. Exp. Biol. Med.* **126**: 830–837, 1967.

Taft, C. H. "Data Processing in the Clinical Laboratory." *Medical Lab.* **1**: 15–18, 1968

APPENDIX

This appendix is by no means to be considered a complete listing of all manufacturers. It is only to provide reference to materials, animals, and equipment with which we have had experience, and, in many cases, referred to specifically in the text of the various chapters. There are other adequate systems available with which we have not personally had experience.

A. Manufacturers

CAGE AND BOTTLE WASHING EQUIPMENT

(1) Better Built Machinery Company
73–75 East 130th St.
New York, N.Y.

(2) Industrial Washing Machine
Corporation
Matawan, N.J.

(3) Specialty Equipment Co.
P.O. Box 4182—North Station
Winston-Salem, N.C.

(4) R. G. Wright Co., Inc.
2280 Niagara St.
Buffalo, N.Y.

BEDDING

(1) Wood Shavings
Black Brook Products Co., Inc.
P.O. Box 45
Lake Port, N.H.

(2) Sterolit
Frederick Gumm Chemical Co., Inc.
538 Forest St.
Kearny, N.J.

(3) San-I-Cel and Pel-E-Cel
Laurel Farms, Inc.
Paxton Processing Co., Inc.
Paxton, Ill.

(4) Absorb-dri
Lab Products, Inc.
635 Midland Ave.
Garfield, N.J.

FEED

(1) Ralston Purina Company
Checkerboard Square
St. Louis, Mo.

(2) Country Best, Agway, Inc.
P.O. Box 148
Waverly, N.Y.

(3) Teklab, Inc.
Monmouth Trust & Savings Bank
Monmouth, Ill.

(4) Nutritional Biochemicals Corp.
Cleveland, Ohio

INCINERATORS

(1) Joseph Goder Company
4241 Honore St.
Chicago, Ill.

(2) Silent Glow Corp.
850 Windsor St.
Hartford, Conn.

CAGING

(1) Hoeltge Bros., Inc.
1919 Gest St.
Cincinnati, Ohio

(2) Lab-Cages, Inc.
126 John St.
Hackensack, N.J.

(3) Maryland Plastics
9 East 37th St.
New York, N.Y.

(4) Wahmann Manufacturing Co.
P.O. Box 6883
Baltimore, Md.

AUTOCLAVES

(1) American Sterilizer Co.
Erie, Pa.

(2) Wilmot-Castle
P.O. Box 629
Rochester, N.Y.

GERMFREE EQUIPMENT

(1) Germfree Laboratories, Inc.
5644 N.W. 7th St.
Miami, Florida

(2) Snyder Manufacturing Co.
New Philadelphia, Ohio

(3) Reyniers and Sons
3806 N. Ashland Ave.
Chicago, Ill. 60613

(4) Lab-Cages, Inc.
126 John St.
Hackensack, N.J.

AUTOMATIC SYRINGE

Becton Dickinson & Co.
Rutherford, N.J.

FACE MASKS

Minnesota Mining & Mfg. Co.
2501 Hudson Rd.
St. Paul, Minn.

AUTOMATIC CHLORINATOR

Wallace & Tiernan, Inc.
25 Main St.
Belleville, N.J.

STOPPER AND SIPPER TUBE WASHER

Ritter Pfaudler Corp.
(R. G. Wright Co. Division)
2280 Niagara St.
Buffalo, N.Y.

RADIOGRAPHIC FILM

(1) Eastman Kodak Co.
Dept. 8-TR
343 State St.
Rochester, N.Y.

(2) General Aniline & Film Corp.
Antco Bldg. at Charles
Binghamton, N.Y.

SUTURE MATERIAL

(1) Ethicon, Inc.
Div. of Johnson & Johnson
Somerville, N.J.

(2) Becton Dickinson & Co.
Rutherford, New Jersey

INSECT SPRAY

Dow Chemical Company
Midland, Michigan

B. Sources of Laboratory Mice

INBRED MICE

(1) Cumberland View Farms
    Clinton, Tenn.
(2) The Jackson Laboratory
    Bar Harbor, Maine
(3) Texas Inbred Mice Co.
    6140 Almeda
    Houston, Tex.

RANDOMBRED MICE

(1) Carworth (Div. of Becton, Dickinson & Co.)
    New City, Rockland County, N.Y.
(2) The Charles River Breeding Laboratories, Inc.
    251 Ballardvale St.
    Wilmington, Mass.
(3) National Laboratory Animal Co.
    P.O. Box 93
    Creve Coeur, Mo.
(4) Manor Research Farms
    Stattsburg, N.Y.

GERMFREE MICE

(1) The Charles River Breeding Laboratories, Inc.
    251 Ballardvale St.
    Wilmington, Mass.
(2) A. R. Schmidt Co., Inc.
    2840 Latham Drive
    Madison, Wisconsin

SPECIFIC-PATHOGEN-FREE MICE

(1) Cumberland View Farms
    Clinton, Tenn.
(2) Manor Research Farms
    Stattsburg, N.Y.

C. Suggested Library

Many of the texts listed here will have been mentioned as recommended reading material at the end of each chapter. It will serve only as a master list as such for texts required for a complete mouse library.

BOOKS AND MONOGRAPHS

Barnes, C. D., and L. G. Eltherington. *Drug Dosage in Laboratory Animals: A Handbook* (Berkeley, Calif.: University of California Press, 1964), 302 pp.

Care and Management of Laboratory Animals, Department of the Army Technical Bulletin: Department of the Air Force Pamphlet, TB Med 255, AFP 160–12–3. May 1958, 112 pp.

Conalty, M. L., ed. *Husbandry of Laboratory Animals: 3rd Symposium of the International Committee on Laboratory Animals* (New York: Academic Press, 1967), 650 pp.

Cook, Margaret J. *The Anatomy of the Laboratory Mouse* (New York: Academic Press, 1965), 143 pp.

Cooper, John A. D. "Foreword," in *A Symposium on . . . Laboratory Animals: Their Care and Their Facilities*, reprinted by U.S. Department of Health, Education, and Welfare, Public Health Service, from *J. Med. Educ.* 35: 55, 1960.

Cotchin, Ernest, and Francis J. C. Roe, eds. *Pathology of Laboratory Rats and Mice* (Philadelphia: F. A. Davis Company, 1967), 848 pp.

Crandall, Lee S. *The Management of Wild Mammals in Captivity* (Chicago: The University of Chicago Press, 1964), 769 pp.

Gay, William I., ed. (a) *Methods of Animal Experimentation*, Vol. I (New York: Academic Press, 1965), 382 pp. (b) *Methods of Animal Experimentation*, Vol. II (New York: Academic Press, 1965), 608 pp.

Green, Earl L. "Introduction," in *Handbook on Genetically Standardized Jax Mice* (Bar Harbor, Maine: Bar Harbor Times Publishing Company, 1962), 83 pp.

Green, Earl L., ed. *Biology of the Laboratory Mouse*, 2nd ed. (New York: McGraw-Hill Book Company, 1966), 706 pp.

Harris, R. J. C., ed. *The Problems of Laboratory Animal Disease* (New York: Academic Press, 1962), 265 pp.

Holdenried, Robert, ed. *Viruses of Laboratory Rodents: Symposium on Viruses of Laboratory Rodents*, Natl. Cancer Inst. Monogr. No. 20 (Washington, D.C.: U.S. Govt. Print. Office, 1965), 180 pp.

Inbred Strains of Mice, No. 5 (Companion Issue to *Mouse News Letter* No. 37), (Bar Harbor, Maine: The Jackson Laboratory, July 1967), 117 pp.

Jones, T. C. "Foreword," in *A Symposium on Some Infectious Diseases of Laboratory Rodents. J. Natl. Cancer Inst.* **20**, No. 5, May 1958. (A Publication of the U.S. Department of Health, Education, and Welfare, Public Health Service, National Institutes of Health, Washington, D.C.)

Jones, T. C., and C. A. Gleiser, eds. *Veterinary Necropsy Procedures* (Philadelphia: J. B. Lippincott Co., 1954), 136 pp.

Kirk, Robert W., ed. *Current Veterinary Therapy: Small Animal Practice, 1966–1967* (Philadelphia: W. B. Saunders Co., 1966), 723 pp.

Klieneberger-Nobel, E. *Pleuropneumonia-Like Organisms (PPLO) Mycoplasmataceae* (New York: Academic Press, 1962), 157 pp.

Laboratory Animals, Part II: Animals for Research. A Report of the Institute of Laboratory Animal Resources: National Academy of Sciences–National Research Council. Publication 1413, Washington, D.C., 1966.

Laboratory Animal Science, A Review of the Literature prepared by Robert J. Flynn. Argonne National Laboratory, ANL–7300, 1st ed., Biology and Medicine (TID–4500), AEC Research and Development Report, June 1966. 58 pp.

Laboratory Animal Science, A Review of the Literature prepared by Robert J. Flynn. Argonne National Laboratory, ANL–7300, 2nd ed., Biology and Medicine (TID–4500), AEC Research and Development Report, November 1966. 133 pp.

Lane-Petter, W., ed. *Animals for Research: Principles of Breeding and Management* (New York: Academic Press, 1963), 531 pp.

Lane-Petter, W., A. N. Worden, F. Hill, J. S. Paterson, H. G. Vevers and the Staff of UFAW, eds. (with a foreword by Sir Peter Medawar). *The UFAW Handbook on the Care and Management of Laboratory Animals*, 3rd ed. (Baltimore: The Williams & Wilkins Co., 1967), 1015 pp.

Loutit, John F. *Irradiation of Mice and Men* (Chicago: The University of Chicago Press, 1962), 154 pp.

Luckey, Thomas D. *Germfree Life and Gnotobiology* (New York: Academic Press, 1963), 512 pp.

Maisin, J. H. "Foreword," in *The Experimental Animal in Research*, Laboratory Animal Science Association Symposium. Reprinted from *Food Cosmetics Toxicol.* **3**, Nos. 1 and 2. (London: Pergamon Press, 1965), 228 pp.

Miyakawa, M., and T. D. Luckey, eds. *Advances in Germfree Research and Gnotobiology* (Cleveland: The Chemical Rubber Company Press, 1968), 439 pp.

Porter, George, and W. Lane-Petter, eds. *Notes for Breeders of Common Laboratory Animals* (New York: Academic Press, 1962), 208 pp.

Riley, William F., Kenneth W. Smith, and Robert J. Flynn, eds. *Year Book of Veterinary Medicine*, Vol. 1 (Chicago: Year Book Medical Publishers, 1963), 456 pp.

―――. *Year Book of Veterinary Medicine*, Vol. 2 (Chicago: Year Book Medical Publishers, 1964), 448 pp.

―――. *Year Book of Veterinary Medicine*, Vol. 3 (Chicago: Year Book Medical Publishers, 1966), 460 pp.

Russell, W. M. S., and R. L. Burch. *The Principles of Humane Experimental Technique* (Springfield, Ill.: Charles C Thomas, Publisher, 1959), 238 pp.

Rugh, Roberts. *The Mouse: Its Reproduction and Development* (Minneapolis: Burgess Publishing Co., 1968), 430 pp.

Schermer, Siegmund. *The Blood Morphology of Laboratory Animals*, 3rd ed. (Philadelphia: F. A. Davis Company, 1967), 200 pp.

Searle, A. G. *Comparative Genetics of Coat Colour in Mammals* (London: Academic Press, 1968), 308 pp.

Short, Douglas J., and Dorothy P. Woodnott, eds. *The A.T.A. Manual of Laboratory Animal Practice & Techniques; The Animal Technicians Association* (London: Crosby Lockwood & Son, Ltd., 1963), 350 pp.

Yager, R. H., and F. J. Judge, eds. *ILAR News:* Institute of Laboratory Animal Resources (National Research Council–National Academy of Sciences–National Academy of Engineering, Washington, D.C., Quarterly Newsletter).

PERIODICALS

Laboratory Animals. Journal of the Laboratory Animal Science Association (London: Laboratory Animals Ltd., Quart. 1968), April 1968.

The Bulletin National Society for Medical Research (no charge). Published monthly at 1530 Massachusetts Ave., N.W., Washington, D.C. 20050.

Laboratory Animal Care. Published by the American Association For Laboratory Animal Science, Joliet, Illinois 60434. Bimonthly publication.

Purina News Notes (no charge). Monthly publication, *Laboratory Animal Digest.* Published by Ralston Purina Co. at St. Louis, Mo.

D. Organizations as Sources of Information

(1) American Association for
 Accreditation of Laboratory
 Animal Care
 4 E. Clinton St.
 P.O. Box 13
 Joliet, Ill. 60434

(2) American Association for
 Laboratory Animal Science
 P.O. Box 10
 Joliet, Ill. 60434

(3) American Society of Laboratory
 Animal Practitioners
 c/o Dr Arthur C. Peters
 505 King Avenue
 Columbus, Ohio 43201

(4) The Association for Gnotobiotics
 c/o Mr Donald M. Robie
 Oak Ridge National Laboratory
 Biology Division
 Oak Ridge, Tenn. 37830

(5) Institute of Laboratory Animal
 Resources
 National Academy of Sciences–
 National Research Council
 2101 Constitution Ave., N.W.
 Washington, D.C. 20418

(6) Laboratory Animal Breeder's
 Association
 6 Beacon St., Suite 626
 Boston, Mass. 02108

(7) American College of Laboratory
 Animal Medicine
 Office of the Secretary-Treasurer
 Institute of Laboratory Animal
 Resources
 2101 Constitution Ave.
 Washington, D.C. 20418

INDEX

A

Acanthocephala (*Moniliformis moniliformis*), 97
Age:
  as a factor in amyloidosis, 61
  in cardiovascular disease, 59
  in hepatic fibrosis, 62
  of mice at maturity, 26, 27
Allergic sensitization to mice, 113–114
Amyloidosis, 61
Anesthetics, 137–143
  electroanesthesia, 143
  inhalation, 142–143
    chloroform, 143
    ether, 140, 142
    halothane, 143
    methoxyflurane (Metofane), 142
  injectable, 137–140
    barbiturates, 138
    narcotics, 139
    other, 139
  resistance to, by SPF mice, 10
Animal caging systems, 17–21
Animal health, 49–114

Animal health (*cont.*):
  degree required for experiments, 49
  diseases, 57–114
    control, 54–57
    diagnosis and therapy (*see* Diagnostic procedures, and specific entries under Noninfectious diseases of mice, Infectious diseases of mice, Parasites of mice, and Nutritional deficiencies, effects of)
    prevention of, 52–54
  effect of personnel on, 55, 56, 74, 75, 96
  nutritional requirements for, 53–54 (*see also* Food; Nutritional deficiencies, effects of)
  testing and monitoring, 50–52 (*see also* Testing and monitoring programs)
Antigen-free environment, 5
Arthropod, human infection from, 112 (*see also* Mites)
Ascites fluid collection, 130–131
Aspergillus flavus, 62
Autoclaves:
  entry portal, 32
  manufacturers of, App. A, 166
Automatic watering systems, 149, 159–161

174 *Index*

B

Bacillus piliformis, in Tyzzer's disease, 65
  confusion with *Salmonella* and *Corynebacterium*, 66
Bacterial diseases (*see* Infectious diseases of mice; Organisms)
Barbiturates, as injectable anesthetics, 138
Barrier systems:
  admittance to, of personnel, 56
  construction and design of, 38–48
    doors, 41
    drains, 42–43
    electrical, 42–43
    sinks, 43
  as disease control for PVM, 80
    for adenovirus infection, 90
  effect on allergy sensitization in humans, 114
  environmental control of, 43–44
  typical facility design, description of, 47
Bartonella infections, 98, 99
Bedding:
  changing of, 26
  dehydration and death from, 24
  disposal of, 25, 42
  manufacturers of, App. A, 165
  materials, 24, 25
  sterilization of, 25
Blood parasites of mice, 98–99 (*see also* Protozoa — blood; Rickettsia)
Blood withdrawal, methods of, 129–130
Breeding (*see also* Conventional mice; Defined-flora mice; Germfree mice; Specific-pathogen-free mice):
  artificial insemination, 137
  commercial breeders, App. B, 168
  computerized data on, 149, 157–159
  effect of PPLO on, 64
  in elimination of Theiler's mouse encephalitis virus, 71
  estrus cycle, 26, 136–137
  gestation period, 27
  maturity, 26, 27
  presence of males after parturition, 27
  systems for, in hybrid mice, 27
    inbred, 27
    randombred, 27
Bronchietosis, enzootic (*see* Catarrh)

C

Cages:
  filter-cap, 28, 53
    systems, 153, 156
  hypoxia chambers, 150–151
  manufacturers of, App. A, 165
  presterilization of, 30
  and racks, 21
  sanitation of, 25, 26
  space per mouse, 26
  types of, 17–21
  washing machines for, 21, 23, 25
Cardiovascular disease in mice, 59
Catarrh, 62–64
  geographical distribution of, 64
  mycoplasma in, 63
  pneumonia virus (CMPV) in, 62
  PPLO in, 62, 63
  Sendai virus in, 62
  Streptobaccilus moniliformis in, 62
Catarrhal enterocolitis, 67
Cestodes (tapeworms), 96
  Coenurus serialis (larva of *Taenia serialis*), 96
  Cysticercus fasciolarus (larva of *T. taeniaformus*), 96
  Hymenolepsis diminuta, 96
  Hymenolepsis nana, 96
  Oocharistica ratti, 96
Chimera, defined, 101
Chloral hydrate, as intravenous or oral anesthetic, 139
Chlorination, 23, 68
Chloroform:
  as anesthetic, 143
  toxicity of, 60, 143
Chronic murine pneumonia virus (CMPV), 64
Chronic respiratory disease (*see* Catarrh)
Citrobacter freundii, in catarrhal enterocolitis, 67
Classification of mice (*see* Ecological classification)
Cockroach (*see also* Pests):
  as carrier of disease, 57
  as carrier of parasite, 97
Computer systems:
  for personnel assignment, 157
  for processing breeding data, 157–159
  purposes of, 156

Computer systems (*cont.*):
  for storage and retrieval of experimental data, 149, 156
Congenital anomaly, in mouse kidney, 61
Contaminants:
  in conventional colonies compared to SPF, GF mice, 10
  in defined-flora mice, 6
  human, 47, 56, 57
  monocontaminants, 5
  polycontaminants, 5
  protection from, by placental barrier, 2
  sources of *Pseudomonas* as, 23
  in SPF colonies, 10
Control:
  of diseases, 54–57
  environmental (*see* Environment)
  of pathogenic parasites, 7 (*see also* entries under Fungi; Rickettsia; Viruses)
Conventional mice, 11–13
  compared to SPF mice, 10
  contamination in, as compared to GF mice, 10
  uses of, 12–13
Corynebacterium kutscheri:
  confusion with *B. piliformis*, 66
  in septic arthritis, 65
Coxsackie virus, A and B groups:
  differentiation of, 81
  infection by, 81
Coxsackie virus infection, 81
Culturing (*see also* entries under specific organism, or Infectious diseases of mice; Parasites of mice; Viruses):
  of LCM, 89
  of MSGV, 83
  of PPLO, 64
  of *Pseudomonas* sp., 68
  of *Salmonella* sp., 69–70

D

Death (*see also* specific diseases listed under Infectious diseases of mice; Noninfectious diseases of mice; Parasites of mice; Viruses):
  from chloroform toxicity, 60, 143
  from chronic respiratory disease, 58
  from dehydration, 24
  from early death syndrome, 52, 58
  from gastric ulcers, 58

Death (*cont.*):
  from hemorrhagic diathesis, 25, 60
  from nutritional deficiencies, 106–109
  from overdose of urethane, 139
  from psychological stress, 58
  from treatment for mites, 97
Defined-flora mice:
  compared to SPF mice, 7
  contaminants in, 7
  defined, 6
  methods of intentional contamination in, 6
Dehydration and death from bedding, 24
Demerol (Meperidene hydrochloride) as injectable anesthetic, 139
Dermatomycotic infections, 90–92
Design:
  barrier facility, 38–48
  construction materials and details, 42
  doors, 41
  drains, 42–43
  electrical, 42–43
  environmental control, 43–44
  human facilities, 57
  overall, 38
  purposes of, 38, 41
Diagnosis, 51 (*see also* specific diseases listed under Infectious diseases of mice; Noninfectious diseases of mice; Parasites of mice; Viruses)
Diagnostic procedures for (*see also* specific diseases listed under Infectious diseases of mice; Noninfectious diseases of mice; Parasites of mice; Viruses):
  Aspicularis tetraptera, 95
  catarrh, 62, 63
  K virus, 73
  LCM, 88
  mouse pox (infectious ectromelia), 85
  MSGV, 84
  PPLO, 79, 80
  Pseudomonas infection, 68
  PVM, 79, 80
  Reovirus type 3, 74, 75
  Salmonella sp., 69
  Syphacia obvelata, 95
  Theiler's mouse encephalitis, 71
  Tyzzer's disease, 66
Diarrhea:
  in catarrhal enterocolitis, 67

Diarrhea (*cont.*):
  in EDIM, 84, 153
  in LIVIM, 84
  in Reovirus type 3 infection, 74
  in Tyzzer's disease, 65, 66
Diet (*see* Food; Nutritional requirements)
Disease control, 54–57 (*see also* specific diseases)
  of K virus, 74
  of LCM, 89
  of mouse pox, 86
  of *Pseudomonas* infection, 68
  in relation to personnel traffic, 55–56
  in relation to pests, 57
  of Reovirus type 3, 76
  of "ringworm", 92
  of Salmonellosis, 69
Disease prevention, 52–54 (*see also* specific diseases)
  in conventional mice, 52–53
  filter-top cages in, 64, 74, 76, 80, 84
  from human sources, 56, 57
  of LCM, 89
  methods, 52–53
  of mouse pox, 87
  of *Pseudomonas* infection, 68
Diseases:
  bacterial, 52 (*see also* Infectious diseases of mice)
  as biomedical models in human disease studies, 59
  cardiovascular, 59
  carried by humans, 56, 74, 75
    by pests, 57, 65, 69, 72
  computer as control aid in, 157
  congenital, 61
  control of (*see* Disease control)
  deep fungal infections, 93
  dermatomycotic infections, 90–92
  diagnosis of, commercial, 51
  diagnosis of, laboratory (*see* specific diseases)
  effect of, on experiments, 58, 61, 72
  encephalitozoonosis, 100
  experimental (*see* Experimental diseases)
  genetic (*see* Genetic diseases)
  infectious (*see* Infectious diseases of mice)
  latent viral, 52, 58
  LCM in humans, 87
  leukemialike, 81
  leukemias, 52, 58
  multiple categories of, 58

Diseases (*cont.*):
  neoplastic, 103–104
    table of, 105–106
  noninfectious (*see* Noninfectious diseases of mice)
  nutritional deficiency (*see* Nutritional deficiencies, effects of)
  parasitic (*see* Parasites of mice)
  prevention of, 52–53 (*see also* Immunization, as concerned with disease prevention; Isolation, in disease prevention; Sanitation; Testing and monitoring programs)
  range of, 52
  rickettsial, 52, 59 (*see also* Eperythrozoon coccoides)
  "ringworm", 90–92
  Sarcosporidiosis, 99–100
  septic arthritis, 65
  specific to mature male mice, 60
  treatment, scale of, 54–55 (*see also* specific diseases)
Drug administration and dosage, 136–137 (*see also* specific diseases)

E

Early death syndrome, 52, 58
  defined, 67
  involvement of *Pseudomonas* in, 67, 68
Ecological classification:
  defined, 1
  determinants of, 16
  effect on zoonotic infections, 113
  of mice:
    conventional, 11–13 (*see also* Conventional mice)
    defined-flora (DF), 6–7
    germfree (GF), 2–5
    specific-pathogen-free mice (SPF), 7–11
Ecological conditions affecting prevalence of Sendai virus infection, 83
Ectromelia, infectious, 84–87
Electroanesthesia, 143
Encephalitis:
  in Reovirus type 3 infection, 74
  Theiler's, 71–72
Encephalomyelitis, infection from laboratory animals, in humans, 113
Entry systems (*see* Isolators)

Environment:
  antigen-free, 5
  in barrier systems, 43
  components of, 1
  control of (*see* Environmental control)
  for conventional mice, 12
  in filter-cage systems, 153, 156
  germfree, reproduction in, 2
  temperature of, 45
Environmental control (*see also*
      Equipment):
  airflow in, 44, 45, 46
  complete system of, described, 44–48
  cooling, 44
  criteria, summary of:
    Table 2.2, 45
  exhaust systems, 46
  as factor in disease, 61
  in filter-cage systems, 153
  heating, 44
  odor, 45
  temperature, 45
Eperythrozoon coccoides:
  association with LDH virus, 81, 99
    mouse hepatitis, 76, 98, 99
  damage to cell function by, 77
  Rickettsial blood parasite, 98
  synergism with *Plasmodium* sp.
      infections, 99
  transmittal by *Polyplax serrata*, 97
Epidermophyton, 91–92
Epizootic diarrhea of infant mice (EDIM),
      84, 153
Equipment:
  alarm systems, 46–47
  cage and bottle washers, 41
  doors, 41
  filter-top cages, 153, 156 (*see also*
      Filter-top cages)
  isolators and entry systems, 31–38
  light control, 43
  manometers, 43
  manufacturers of, App. A, 165–167
  rack washers, 42
  sinks and drains, 42, 43
  sterilizer, 41–42
  thermostats, 43
  X-ray machines, 47–48
Estrus in the mouse, 26–27
  induction of, 136–137
Ether, as inhalation anesthetic, 140

Ethyl carbamate (Urethane) as injectable
      anesthetic, 139–140
Euthanasia, 143–144
Experimental diseases of mice:
  Mycobacterium leprae infection, 101
  secondary disease, 101–102
Experiments:
  complicated by:
    LCM, 90, 113
    lymphomas, 81
    mouse hepatitis, 76
  as complicating factor in:
    K virus infection, 72
    Tyzzer's disease, 61
  computerized programs and analysis of,
      149, 156–158
  on oxygen–carbon dioxide determination,
      151–152
  using roentgenographic techniques and
      radioisotopes, 144–145

F

Facility design (*see* Housing)
Filter-top cages:
  in disease prevention, 64, 74, 76, 80, 84
  effect on allergic-sensitization in
      personnel, 114
  in parasite prevention, 97
  role in concept of pathogen-free
      animals, 149
  systems, 153, 156
  in transfer procedures, 9
Flora indigenous in mice, 7
Fluid:
  administration to mice, 127–129
    intra-arterial, 129
    intramuscular, 127–128
    intraperitoneal, 127
    intrathoracic, 128
    intravenous, 128
    oral, 129
    subcutaneous, 128
  withdrawal from mice, 129–131
    of ascites fluid, 130–131
    of blood, 129–130
    of lymph, 130
    of urine, 131
Food (*see also* Nutritional requirements of
      mice; Nutritional deficiencies, effects
      of):
  amount required (daily consumption), 50

178   *Index*

Food (*cont.*):
  chlortetracycline in, as Leptospirosis control, 69
  effect on physiological behavior, 54
  forms of, 54
  human, as source of contamination, 57
  manufacturers of, App. A, 165
  relation to mouse strain, 54
  role in resistance to disease, 54
  special formulas, 53–54
  sterilization of, 54
  varieties, 54
Foster-nurse, germfree, 3, 4 (*see also* Hysterectomy)
  in elimination of Theiler's mouse encephalitis, 71
Fungi causing dermatomycotic infections:
  Epidermophyton, 91
  Microsporum sp., 91
  Microsporum gypseum, 92
  Trichophyton sp., 91
  Trichophyton mentagrophytes, 92
  Trichophyton quinckeanum, 92

G

Genetic diseases of mice, mutants and hereditary abnormalities, 102–103
Germfree equipment, manufacturers of, App. A, 166 (*see also* Sterilization)
Germfree mice, 2–5 (*see also* Hysterectomy)
  sources of, App. B, 168
Gestation period in mice, 2, 51
Gnotobiotic mice, defined, 5
Grafting, surgical, 133, 136

H

Halothane, as inhalation anesthetic, 143
Hemorrhagic diathesis, 25, 60
Hepatic fibrosis, 62
Hepatitis, 76–78 (*see also* Murine, hepatitis virus)
Hereditary abnormalities, 102–103
Hexobarbitol, as injectable anesthetic, 138
Housing (*see also* Isolators):
  cages, 17
  of conventional mice, 28
  location of, 28

Housing (*cont.*):
  as source of infection, 77
  space per mouse, 26
  of SPF, DF, GF mice, 29
  techniques, 31, 47
  type of, effect on allergic sensitization of personnel, 114
Human infectability (*see also* Personnel):
  by encephalomyelitis, 113
  by LCM, 87
  from mites, 112
  by murine hepatitis virus, 112
  by "ringworm," 92
Hybrid, tracing lineage of, 158
Hypervolemia, 61
Hypoxia chambers, 150–153
Hysterectomy, 2, 3, 10 (*see also* Surgery)
  disease prevention through derivation by, 64, 74, 76, 80, 84, 90

I

Immunity:
  to catarrh, 64
  to LCM, 88
  to *Trichophyton*, 92
Immunization, as concerned with disease prevention, 53
Inbred mice (*see also* Defined-flora mice; Germfree mice; Specific-pathogen-free mice):
  chloroform toxicity in, 60
  first strain of, 6
  hepatic fibrosis in, 62
  kidney disease in, 61
  lymphomas in, 81
  sources of, App. B, 168
  tracing lineage of, 158
Inbreeding systems, 27
Incinerators, manufacturers of, App. A, 165
Infantile diarrhea (EDIM, LIVIM), 84, 153
Infarction, 61
Infectious catarrh (*see* Catarrh)
Infectious diseases of mice, 62–95
  Bartonella infections, 99
  catarrh, 62–64
  catarrhal enterocolitis, 67
  Coxsackie virus infection, 81
  deep fungal infections, 93
  dermatomycotic infections ("ringworm"), 90–92

Infectious diseases of mice (*cont.*):
   hepatitis, 76–78 (*see also* Murine, hepatitis virus)
   infantile diarrhea (EDIM, LIVIM), 84
   infectious ectromelia (mouse pox), 84–87
   K virus infection, 72–74
   lactic dehydrogenase virus (LDH) infection, 81, 99
   leprosy, 67
   leptospirosis, 69
   lymphocytic choriomeningitis (LCM), 57, 87–90
   lymphomas, 81
   mouse adenovirus infection, 90
   mouse encephalomyelitis, Theiler's, 57, 71–72
   mouse pox (infectious ectromelia), 84–87
   mouse salivary gland virus infection, 83–84
   mouse typhoid (Salmonellosis), 69–70
   mycoplasmal infections, miscellaneous, 64
   Pasteurella infection, 70
   Plasmodium sp. infections, 99
   pneumonia virus infection, 79–80 (*see also* Pneumonitis)
   polyoma virus infection, 78–79
   Pseudomonas infection, 67–68
   Reovirus type 3 infection, 74–76
   "ringworm," 90–92
   salmonellosis (mouse typhoid), 69–70
   Sendai virus infection, 82–83
   septic arthritis, 65
   Theiler's mouse encephalomyelitis, 71–72
   tularemia, 71
   Tyzzer's disease, 65–67
Information on laboratory animals and their care, organizations as sources of, App. D, 172
Irradiated mice:
   Pseudomonas in, 67
   Tyzzer's disease in, 65
Irradiation:
   in experimental disease, 101–102
   special techniques involving, 131, 133
Isolation, in disease prevention, 53, 64
Isolators:
   cages, 38, 53
   in production of GF mice, 2, 3
   types of, and entry systems, 31–38
      Table 2.1, 33

K

Kidney disease, 61
K virus:
   description of, 72
   in K virus infection, 72–74
   in various classifications of mice, compared, 10
K virus infection, 72–74

L

Leptospira ballum, 69
Leptospirosis, 69
Lethal intestinal virus of infant mice (LIVIM), 84
Leukemialike disease in mice, 81–82
Leukosis, 57
Literature on the laboratory animal, master list, 169–171
Little, Clarence Cook, 6
Lymphocytic choriomeningitis, 57, 87–90
   human infection from, 113
Lymphoma, 81
Lymph withdrawal from mice, 130

M

Maintenance procedures, 16
Manufacturers of equipment, materials, and food, App. A, 165–167
Mating, 27
   age at, 51
   season, 51
   use of computer programming in, 158
Meperidine hydrochloride (Demerol), as injectable anesthetic, 139
Methoxyflurane (Metofane), as inhalation anesthetic, 142
Mice:
   basic data on, Table 3.1, 50
   conventional, 11–13
   defined-flora, 6–7
   germfree, 2–5
   gnotobiotic, 5
   sources of, App. B, 168
   specific-pathogen-free, 7–11
Microsporum, 91
   Microsporum gypseum, 92

Miscellaneous:
  noninfectious diseases, *Aspergillus flavus* in, 62
  parasites and larva, 100
Mites, 97, 98
  Myobia musculi, 97
  Myocoptes musculinus, 97
  Myocoptes romboutsi, 97
  Notoedres muris, 97
  Ornithonyssuc bacoti, 97
  Polyplax serrata (louse), 97, 98
  Polyplax spinulosa, 98
  Psorergates simplex, 97
  Radfordia affinis, 97
Monitoring programs (*see* Testing and monitoring programs)
Monocontaminant gnotobiotes, 5
Morphine, as injectable anesthetic, 139
Mouse:
  adenovirus, 90
    in various classifications of mice, compared, 10
  encephalomyelitis, Theiler's, 71–72
    cockroach as carrier, 57
  poliomyelitis (*see* Theiler's mouse encephalomyelitis)
  pox (infectious ectromelia), 84–87
  salivary gland virus, 83–84
    in various classifications of mice, compared, 10
  typhoid (Salmonellosis), 69–70
Murine:
  hepatitis virus, 76–77
    E. coccoides synergism in, 98–99
    as experimental complication, 76
    human infection from, 112
    induced by stress, 76
    in various classifications of mice, compared, 10
  pneumonotropic viruses, 62, 79–80
    associated with lymphomas, 81
    chronic murine pneumonia virus (CMPV), 64
    K virus, 73
    spontaneous infection with, 64
Mutants, 102–103
Mycobacterium leprae murium, 67
  in experimental disease, 101
Mycoplasma:
  in catarrh, 63
  miscellaneous infections from, 64

Mycoplasma (*cont.*):
  Mycoplasma arthriditis in septic arthritis, 65
Myxoviruses, 76

N

Narcotics, as injectable anesthetics, 139
Nematodes, 95–96
  Aspicularis tetraptera, 95
  Heterakis spumosa, 95
  Nippostrongyles muris, 96
  Syphacia obvelata, 95, 96
Neoplastic disease, 103–104
  table of common tumors, 104–105
Noninfectious diseases of mice, 59–62
  amyloidosis, 61
  chloroform toxicity, 60
  hemorrhagic diathesis, 60
  hepatic fibrosis, 62
  hypervolemia, 61
  infarction, 61
  kidney diseases, 61
  miscellaneous, 62
  ringtail-like disease, 61–62
  spontaneous polyarteritis, 59
Nutritional deficiencies, effects of (nutritional diseases), 106–110
  carbohydrates, 107
  fats, 106–107
  fat-soluble vitamins, 108
  minerals, 107–108
  protein, 106
  water-soluble vitamins, 108–110
Nutritional requirements of mice, 53–54 (*see also* Food)
  effects of sterilization on, 54
  relation of diet to mouse strain, 54
  table of, 55–56

O

Organisms:
  Aspergillus flavus, 62
  Bacillus piliformis, 65
  Citrobacter freundii, 67
  Corynebacterium kutscheri, 65
  Eperythrozoon coccoides, 76, 77, 98
  Leptospira ballum, 69
  Mycobacterium leprae murium, 67

Organisms (*cont.*):
  Mycoplasma, 63
    Mycoplasma arthriditis, 65
    Pasteurella pneumonotropica, 70
    Pasteurella tularensis, 71
  in personnel, contaminating, 56 (*see also* Personnel)
  PPLO (pleuro-pneumonialike-organisms), 62
  Pseudomonas sp., 23, 47, 52, 67, 68, 69
    Pseudomonas aeruginosa, 13, 58
    Pseudomonas stutzeri, 68
  Streptobacillus enteriditis, 69
    Streptobacillus moniliformis, 62, 64
    Streptobacillus typhimurium, 69
Organizations, as sources of information on laboratory animal care, App. D, 172

P

Parasites of mice, 95–100
  blood, 98–99 (*see also* Protozoa — blood; Rickettsia)
  external, 97 (*see also* Mites)
  intestinal, 95–97 (*see also* Acanthocephala; Cestodes; Nematodes; Protozoa — intestinal)
  miscellaneous, 100
    larval form, 100
  other, by organ, 99–100 (*see also* Protozoa — other)
  pathogenic, control of, 7
Pasteurella infection, 70
Pasteurella pneumotropica, 70
Pasteurella tularensis, 71
"Pathogen-free" mice, 7
  contaminants in as compared to conventional mice, 10
  in filter-cage systems, 156
Pentobarbitol, as injectable anesthetic, 138
Permselective membranes, 151–152
Personnel:
  allergic sensitization to mice, 113–114
  clothing of, 56, 156
  and computer systems, 158–159
  entry to facilities, 47, 56
  health of, 56
  infected by LCM, 87
  infected by "ringworm," 92
  infections in, 56
  responsibility of, 28

Personnel (*cont.*):
  traffic, effect on colony health of, 55, 56, 74, 75, 96
  zoonoses in, 110–113
Pests:
  control of, 57
  defined, 57
  in relation to specific diseases, 65, 69, 72, 83, 84, 87, 89, 92
Placenta, 3
Placental barrier, 4, 10
  anesthetic capable of crossing, 142
  as protection against contamination, 2
  to PVM, 80
  viruses capable of crossing, 10
Plasmodium sp. infections, 99
Pneumonia, chronic murine (*see* Catarrh)
Pneumonia virus (*see* Murine, pneumonotropic viruses)
Pneumonitis, 79–80
  Pasteurella pneumonotropica in, 70
  produced by K virus, 73
    by PVM, 79–80
    by Sendai virus, 62
  Reovirus type 3 in, 74
Polycontaminant gnotobiotes, 5
Polycythemia, 152
Polycythemic mice, 150, 152
Polyoma virus, 78–79
  in various classifications of mice, compared, 10
Polyoma virus infection, 78–79
Poxvirus muris, 84
PPLO (pleuro-pneumonialike organisms)
  in catarrh, 62, 63
  miscellaneous infections from, 64
Prophylactic medication, 53
Protozoa — blood:
  Babesia (Piroplasma) muris, 98
  Hemobartonella muris, 98
  Hepatozoon muris, 97, 98
  Plasmodium berghei, 98
  Plasmodium chaubaudi, 99
  Plasmodium vinckei, 98, 99
  Toxoplasma goudii, 97, 98
  Trypanosoma cruzi, 98
  Trypanosoma lewisi, 98
Protozoa — intestinal:
  Chilomastix bettencourti, 97
  Cryptospiridin muris, 97
  Eimeria sp., 97
  Entamoeba muris, 97

Protozoa — intestinal (*cont.*):
  Giardia muris, 97
  Hepatozoon muris, 97, 98
  Hexamita muris, 97
  Klossniella muris, 97
  Toxoplasma gondii, 97, 98
  Trichomonas muris, 97
  Trypanosoma duttoni, 97
Protozoa — other
  Nosema cuniculi (*Encephalitozoon cuniculi*), 100
  Sarcosporidia, 99
Pseudomonas sp.:
  culture of, 68
  infection by, 67, 68
  optimum conditions for, 68
  Pseudomonas aeruginosa, 13, 58
  Pseudomonas stutzeri, 68

Q

Quarantine:
  in mouse pox, 86
  need for, 12
  testing and monitoring in, 51

R

Racks, cage, 21, 23
Radioisotopes, used in experiments with mice, 145
Randombred mice, 27
  sources of, App. B, 168
References, master list of, App. C., 169–171
Reovirus type, 3, 74–76
Reovirus type 3 infection, 74–76
  in humans, 74, 75, 112
Reproduction, in a germfree environment, 2
Restraint of mice:
  difficulty in electroanesthesia, 143
  methods of, 127, 128
Rickettsia (*see also Eperythrozoon coccoides*)
  Hemobartonella muris, 98
Rickettsialike agent, 59
Ringtail-like disease, 61–62
"Ringworm," 90–92
Roentgenographic techniques used with mice, 144–145

S

Salmonella sp., 69, 70, 110
  confusion with *B. piliformis*, 66
  Salmonella enteritidis, 69
  Salmonella newport, 110
  Salmonella typhimurum, 69
    human infection from, 112
Salmonellosis (mouse typhoid), 69–70
Sanitation (*see also* Sterilization):
  of cage facility, 25
  in disease prevention, 53
  effect on allergic sensitization to mice, 114
  level of *Pseudomonas* sp. as indicator of, 67
  personnel, 47, 56, 57
  in pest prevention, 57
  typical system of, in barrier facility, 47
Sarcosporidia, 99
Selective permeability, defined, 151
Sendai virus:
  cause of pneumonitis, 62
  effect of ecological conditions on, 83
  human infection by, 82, 112
  isolation of, 83
  in various murine classifications, compared, 10
Sendai virus infection, 82–83
Septic arthritis, 65
Snuffles (*see* Catarrh)
Sources of laboratory mice, App. B, 168
Specific-pathogen-free mice, 7–11
  in aging studies, 10
  compared to conventional mice, 10
  compared to DF mice, 7
  defined, 7
  in eradication of PVM, 80
  to establish a colony of, 8
  life span, 10
  monitoring a colony of, 10
  procedures and equipment for experiments on, 48
  resistance to anesthetics of, 10
  sources of, App. B, 168
Spontaneous polyarteritis, 59
Sterilization:
  of air, 30
  by autoclave, 24
  of bedding, 25, 47
  of cages, 23
    by gas, steam, liquid chemicals, 30

Sterilization (*cont.*):
  of clothing, 47
  effect of, on nutrient value of food, 54
  by ethylene glycol, 25
  of food, 47
  of miscellaneous equipment, 47
  of racks, 21, 47
  by steam, 23, 25
  of watering equipment (bottles, stoppers, tubes), 21, 23, 24, 47
Streptobacillus moniliformis:
  in catarrh, 62
  in septic arteritis, 64
Surgery:
  hysterectomy, 2, 3
  thymectomy, as experimental parameter, 101
Surgical grafting, 133, 136
Surgical procedures, 125–127
Symptoms (*see* specific diseases)

T

Tables:
  2.1 Basic isolators and entry systems, 33
  2.2 Summary of environmental criteria, 45
  3.1 Basic data on the common laboratory mouse, *Mus musculus*, 50–51
  3.2 Generally accepted nutritional requirements for the laboratory mouse, 55–56
  3.3 Incidence of common tumors in various inbred strains of the laboratory mouse, 104–105
Tapeworms (*see* Cestodes)
Techniques:
  of anesthesia (*see* Anesthetics)
  defined, 125
  for derivation of GF mice, 2, 3
  of drug administration and dosage, 136–137 (*see also* Anesthetics)
  of euthanasia, 143, 144
  of fluid administration and removal (*see* Fluid)
  of radioisotope application, 145
  of restraint, 127, 128
  roentgenographic, 144–145
  special:
    involving shielding and irradiation, 131, 133

Techniques (*cont.*):
  of surgical grafting, 133, 136
  surgical, 125–127
  of vaccination, 86
Testing and monitoring programs:
  adjustment to research purpose, 12
  for closed colonies, 52
  commercial, 51
  importance in LCM control, 89
  lymphomas in, 81
  for mice from commercial sources, 51–52
  need and purpose of, 49, 50
  random, 52
  range of, 52
  routine, for pathogens, 52
  spot-sampling in, 52
  test-and-slaughter method, 53
Theiler's mouse encephalomyelitis, 71–72
  cockroach as source of, 57
  comparative occurrence in various classifications of mice, 10
Theiler's mouse encephalomyelitis virus:
  human infection from, 112
  infection in mice, 71–72
Therapy (*see* specific diseases)
Thiopental, as injectable anesthetic, 138
Thymectomy, 101
Thymic agent, in various classifications of mice, compared, 10
Treatment (*see* specific diseases)
Tribromethanol, as injectable anesthetic, 139
Tularemia, 71
Tumors, spontaneous, 104
  table of common tumors, 104–105
Tyzzer's disease, 65–67

U

Urethane (ethyl carbamate), as injectable anesthetic, 139–140
Urine collection, 131

V

Vaccination, against mouse pox, 86
Vertical transmission of viruses, 4
Virus:
  capable of crossing placental barrier, 10

Virus (*cont.*):
  connection with lymphomas, 81
  in conventional mice compared to SPF mice, 10
  incidence of, 10
  order of prevalence of, 82
  transmittal of, 4
Viruses:
  CMPV, 64
  Coxsackie virus, 81
  EDIM, 84, 153
  K virus, 72, 73, 74
  LCM, 87, 88, 89, 90
  LDH virus, 81, 99
  LIVIM, 84
  mouse hepatitis viruses, 76, 98–99
  MSGV, 83
  polyoma virus, 78, 79
  Poxvirus muris, 84
  PVM, 62, 79–80
  Reovirus type 3, 74, 75, 76
  Sendai virus, 62, 82, 83
  Theiler's mouse encephalomyelitis virus, 71

Vitamins:
  destruction of, 54
  required, table of, 56

W

Washing machines:
  for bottles, 21, 23
  for cages, 21, 23, 25
Water:
  amount required (daily consumption), 50
  changing of, frequency, 26, 47
  chlorination of, 23, 46, 68
  Terramycin in, 66
Watering systems:
  automatic, 149, 159–161
  bottles for, 21
  as source of contamination, 23, 47

Z

Zoonoses, 110–113